Mechanical Desktop® 5 Update Guide

Mechanical Desktop® 5 Update Guide

ED O'HALLORAN
DANIEL T. BANACH

autodesk® press

THOMSON LEARNING™

Australia • Canada • Mexico • Singapore • Spain • United Kingdom • United States

autodesk®
press

Mechanical Desktop® 5 Update Guide

Ed O'Halloran / Daniel T. Banach

Autodesk Press Staff

Business Unit Director:
Alar Elken

Executive Editor:
Sandy Clark

Acquisitions Editor:
James DeVoe

Developmental Editor:
John Fisher

Editorial Assistant:
Jasmine Hartman

Executive Marketing Manager:
Maura Theriault

Marketing Coordinator:
Karen Smith

Executive Production Manager:
Mary Ellen Black

Production Manager:
Larry Main

Production Editor:
Stacy Masucci

Art and Design Coordinator:
Mary Beth Vought

Cover Image:
Matt McElligott

Printed in Canada
1 2 3 4 5 XXX 06 05 04 02 01

For more information, contact Autodesk Press, 3 Columbia Circle, PO Box 15015, Albany, New York, 12212-5015.

Or find us on the World Wide Web at www.autodeskpress.com

Library of Congress Cataloging-in-Publication Data
O'Halloran, Ed
Mechanical Desktop 5 Update Guide /
Ed O'Halloran, Daniel T. Banach.
p. cm.
ISBN 0-7668-2806-9
1. Civil engineering–Computer programs. 2. Surveying–Computer programs. 3. AutoCAD.
I. Title.
TA345 .Z54 2000
624'.0285'5369—dc21

Notice To The Reader

The publisher does not warrant or guarantee any of the products described herein or perform any independent analysis in connection with any of the product information contained herein. The publisher does not assume and expressly disclaim any obligation to obtain and include information other than that provided to it by the manufacturer.

The reader is expressly warned to consider and adopt all safety precautions that might be indicated by the activities described herein and to avoid all potential hazards. By following the instructions contained herein, the reader willingly assumes all risks in connection with such instructions.

The publisher makes no representations or warranties of any kind, including but not limited to, the warranties of fitness for particular purpose or merchantibility, nor are any such representations implied with respect to the material set forth herein, and the publisher and author take no responsibility with respect to such material. The publisher shall not be liable for any special, consequential, or exemplary damages resulting, in whole or in part, from the readers' use of, or reliance upon, this material.

CONTENTS

INTRODUCTION

Welcome! If you are new to Mechanical Desktop or 3D design, you have just joined over 203,000 people already using Mechanical Desktop. If you are a current Mechanical Desktop user, you will find major enhancements in the software over the previous release.

The chapters in this book follow the order in which you will create your own models and drawings. Each chapter introduces a set of topics and then provides detailed explanations. Each chapter builds on the material learned in the previous chapter(s). At the end of most chapters you will find tutorials for you to complete on your own. They are based on real world parts used in different disciplines of design.

PRODUCT BACKGROUND

Mechanical Desktop 5 was written by Autodesk and runs inside of AutoCAD 2000i. Mechanical Desktop is a 3D feature-based parametric solid modeler that allows you to create complex 3D parametric models and to generate 2D views from those models.

Mechanical Desktop 5 consists of AutoCAD 2000i and five modules:

Designer: Feature-based parametric modeler. Part Modeling.

AutoSurf: Non Uniform Rational B-Splines (NURBS) surfaces. Surface Modeling.

Assembly: Manage and constrain assembled parts. Assembly Modeling.

Drawing Manager: 2D view layout and dimensioning for outputting engineering drawings.

Power Pack: 2D and 3D Standard Parts database and productivity tools.

REQUIREMENTS

This book assumes that you are running Mechanical Desktop 5 and that you are proficient with AutoCAD commands such as lines, arcs, circles, polylines, move, erase, grips etc. If you are not proficient in those areas, you may want to refer to the AutoCAD online help as needed.

BASICS OF 3D MODELING

If you are new to creating 3D models, you need to take time to evaluate what you are going to model and how you are going to approach it. When I evaluate a part, I look for the main basic shape. Is it flat or cylindrical in shape? Depending on the shape, I will take a different approach to the model. I try to start with a flat face if possible; it is easier to add other features to a flat face. If the model is cylindrical in shape, I look for the main profile or shape of the part and revolve or extrude that profile. After the main body is created, work on the other features and look to see how this shape will connect to the first part. Think of 3D modeling as working with building blocks: each block sits on another block, but remember that material can also be removed from the original solid.

TERMS AND PHRASES

To help you to better understand Mechanical Desktop, a few of the terms and phrases that will be used in the book are explained below.

Parametric Modeling: Parametric modeling is the ability to drive the size of the geometry by dimensions. For example, if you want to increase the length of a plate from 5" to 6", change the 5" dimension to 6" and the geometry will update. Think of it as the geometry along for a ride, driven by the dimensions. This is opposite to AutoCAD 2D dimensioning, known as associative dimensioning: as lines, arcs and circles are drawn, they are created to the exact length or size; when they are dimensioned, the dimension reflects the exact value of the geometry. If you want to change the size of the geometry, you stretch the geometry and the dimension automatically gets updated. Think of this as the dimension along for a ride, driven by the geometry.

Feature-based parametric modeling: Feature-based means that as you create your model, each hole, fillet, chamfer, extrusion, etc., is an independent feature that can be edited or deleted.

Bi-directional Associativity: The model and the drawing views are linked. If the model changes, the drawing views will automatically update. And if the dimensions in a drawing view change, the model is updated and the drawing views are updated based on the updated part.

OVERVIEW OF PART CREATION

To get a better idea of the process that you will go through to create a part and its 2D views, refer to the steps outlined below. This is intended as an overview only; not all of the steps are required for every feature that is created.

1. Sketch the geometry using lines, arcs, polylines or splines.

2. Profile the sketch. This analyzes the profile, and tells you how many dimensions or constraints are required to fully constrain the profile.

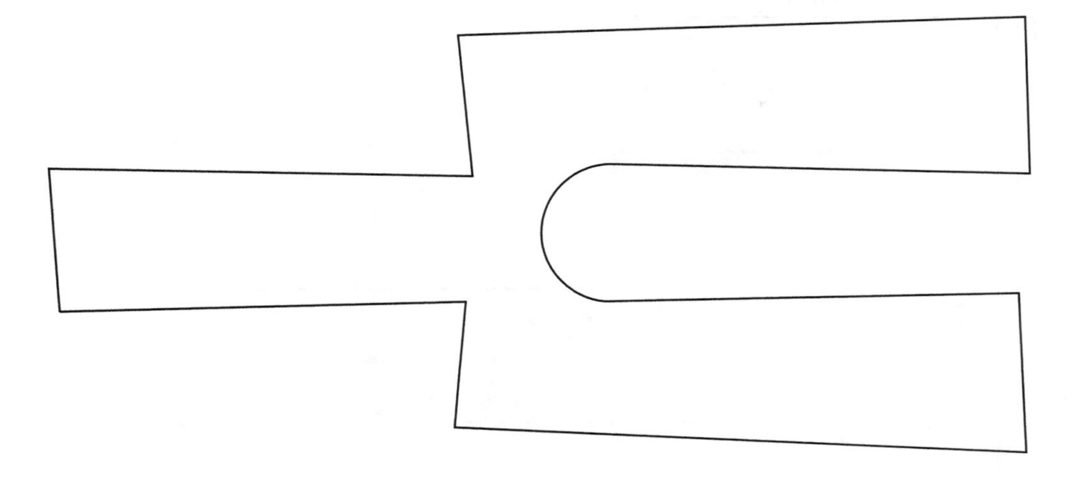

Figure 1

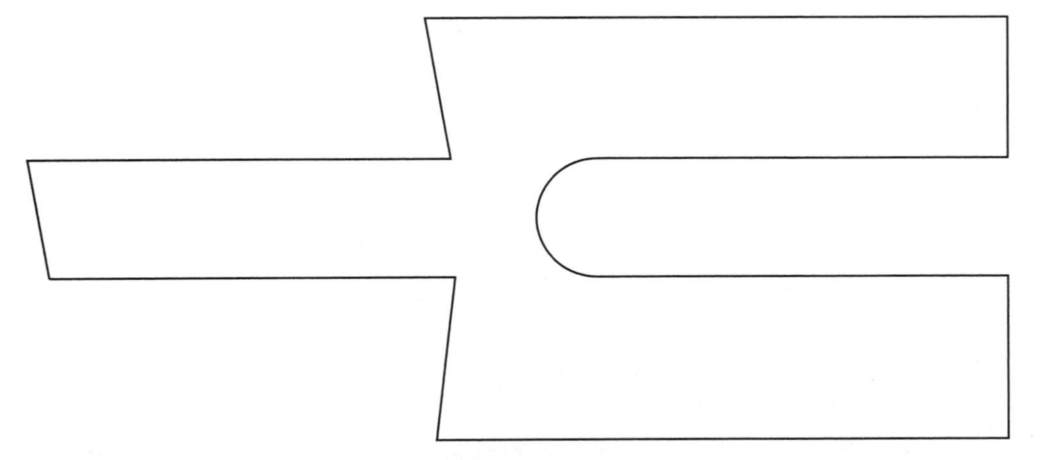

Figure 2

3. Add/remove constraints to control the behavior of the profile.
4. Add dimensions to control the size of the profile.

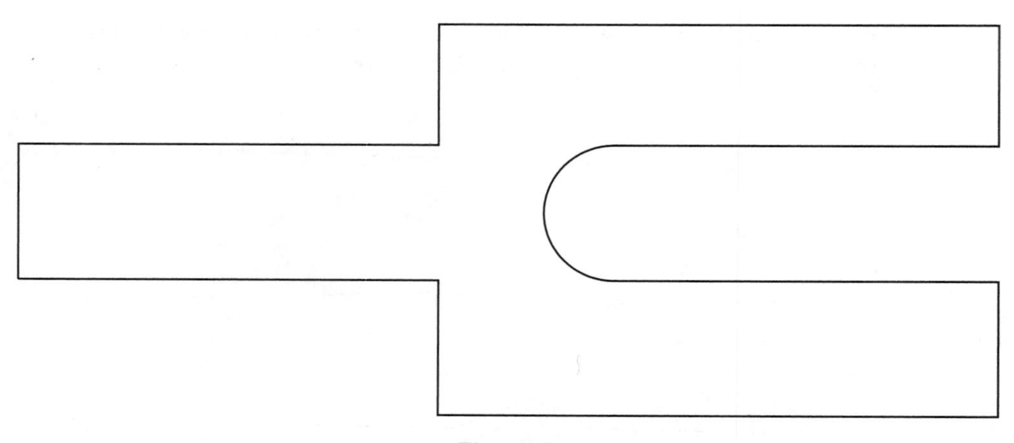

Figure 3

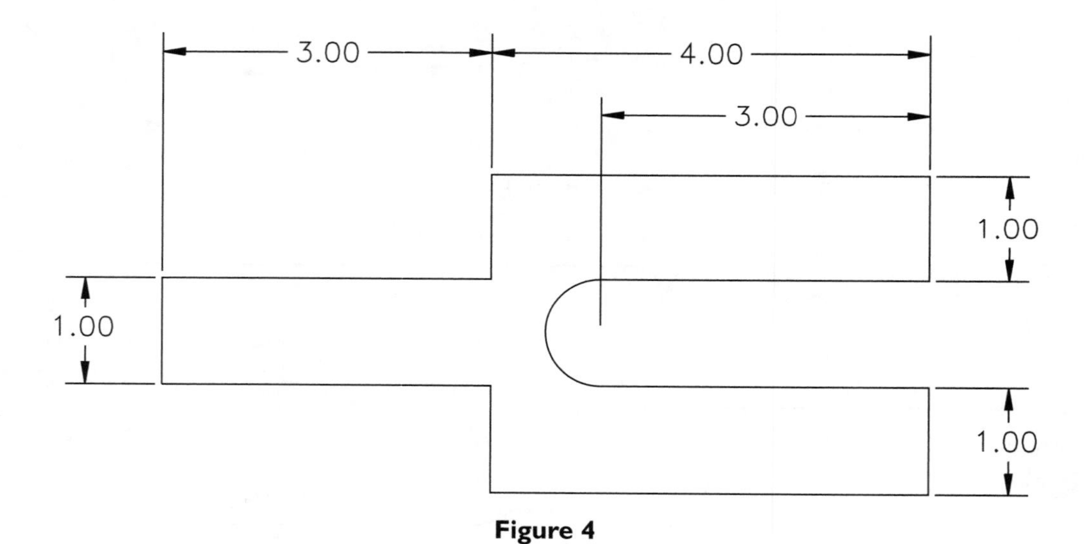

Figure 4

5. Extrude, revolve or sweep the profile into a solid. In Figure 5 the profile was extruded.
6. Add sketched features and placed features such as extrusions, holes, fillets and chamfers.
7. Create the 2D views and annotate the drawing.

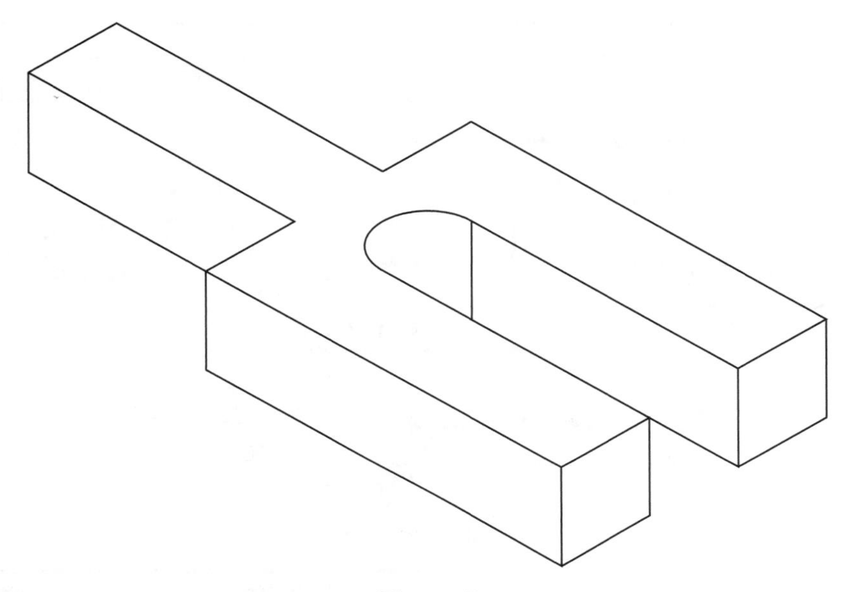

Figure 5

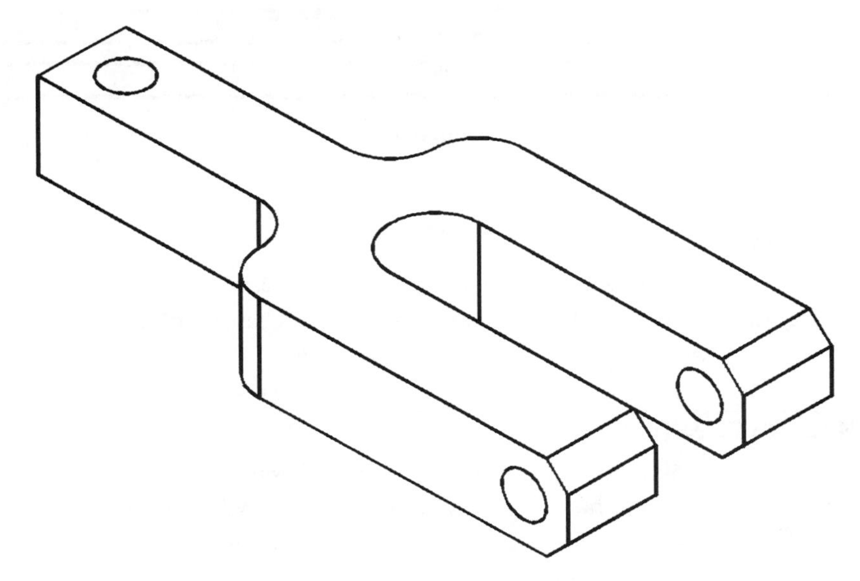

Figure 6

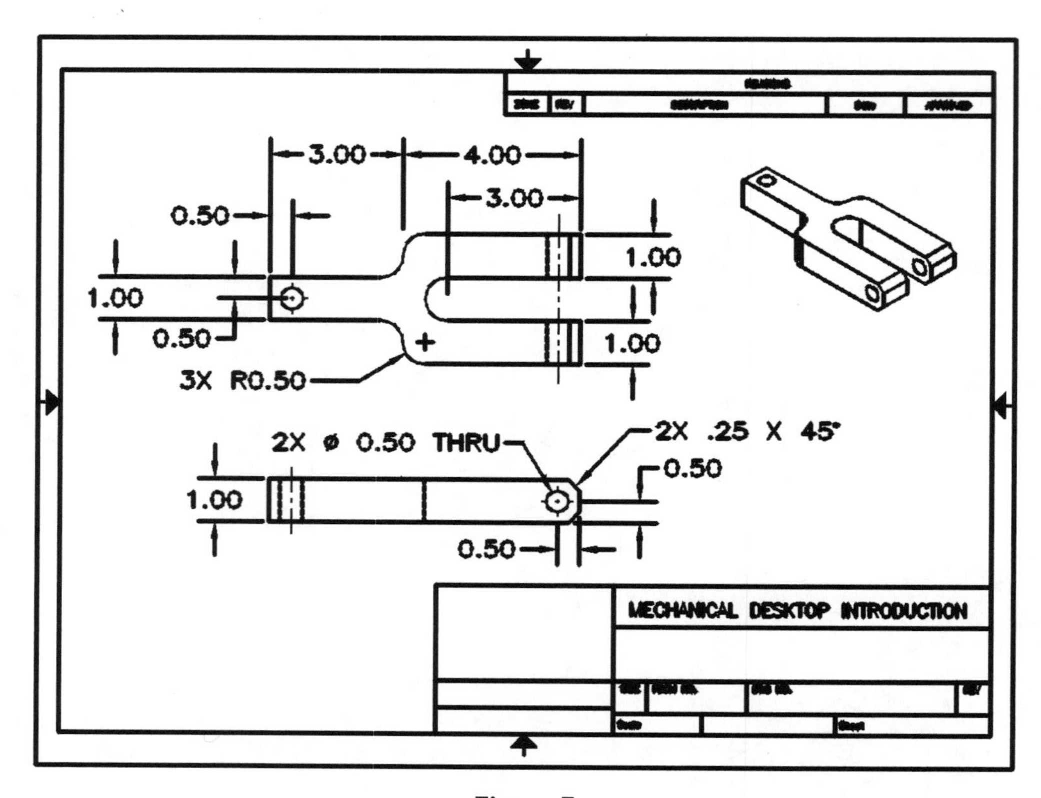

Figure 7

TOOLBARS

In this book, the toolbars will be shown in the default vertical orientation, but remember that this is a personal choice, and the toolbars can be placed and oriented anywhere you want.

BOOK'S INTENT

The intent of this book is to focus on the Designer, Assembly, and Drawing Manager modules in an applied, hands-on environment. Also, Web and Power Pack capabilities of Mechanical Desktop 5 are introduced. This book will guide you through the new and enhanced commands.

Read through each subject and then complete its tutorial while at the computer. At the end of each chapter, there are one or more tutorials you can complete on your own and review questions to reinforce the topics covered in that chapter.

SPECIAL SECTIONS

You will find sections marked:

Note: Here you will find information that points out specific areas that will help you learn Mechanical Desktop.

Tip: Here you will find information that will assist you in generating better models.

BOOK NOTATIONS

- ENTER refers to Enter on the keyboard or right mouse click.
- Numbers in quotation marks are numbers that need to be typed in.
- Part and model both refer to a Mechanical Desktop part.
- Select refers to a pick by the left mouse button.
- Choose refers to a selection from a menu.
- Desktop Browser and Browser both refer to the Mechanical Desktop browser, where the history of the file is shown.
- Sketch refers to lines, arcs, circles and polylines drawn to define an outline shape of a feature.
- Profile refers to a sketch that has been profiled (analyzed) by Mechanical Desktop.

TUTORIAL EXERCISES

The best way to learn Mechanical Desktop is to practice the tutorials and exercises on the computer. Each book ships with a CD with drawings directories for each chapter.

Tip: Make a copy of your Mechanical Desktop 5 startup icon and change the properties of the "Start in:" directory to "c:\MD5Book\chapter??" (where "c" represents the disk drive where the files are located on your system and the "??" represents the chapter you are working on). If you update this startup location as you work through the book, you will reduce the number of picks required to open and save files.

As you go through the book you will find some of the tutorials have already been started for you; they can be opened as noted in each tutorial. As you go through each exercise, save the files as noted, because they may be required in a future exercise.

For clarity, some figures will be shown with lines hidden and the text larger.

With AutoCAD, there are many ways to complete a part, and that is also true with Mechanical Desktop. After completing the exercises as noted, feel free to experiment with different methods.

FILES INCLUDED ON THE CD

You will find all the required drawings on the CD included with this book. Copy the files to your hard drive and remove the read-only property from all the files. It is suggested that you copy the files to C:\MD5Book.

ACKNOWLEDGMENTS

I would like to thank Brian Schanen of Mastergraphics, Wakeshaw, WI for performing the technical edit on the manuscript. I would also like to thank Linda Romanenko for sharing her expertise and furnishing the foundation.

A special acknowledgment is due the following instructors, who reviewed the chapters in detail:

Jeffery R.Gibbs
Muskingum Area Technical College
Zanesville, OH

Steven Keith
De Anza
Cupertino, CA

Gary Masciadrelli
Springfield Technical Community College
Springfield, MA

David A. Probst
Pennsylvania College of Technology
Williamsport, PA

David W. Smith
Cincinnati State Technical & Community College
Cincinnati, OH

Thomas White
Shenendehowa Central School
Clifton Park, NY

ABOUT THE AUTHOR

Ed O'Halloran is a lead instructor and director of Quantum Training Centers Cleveland, OH training site. He has been working with AutoCAD since the mid 80's (version 2.12) and Mechanical Desktop since it's inception. Ed's experience ranges from design and manufacture of race cars and automatic transaxles, to machine and plant layouts in the foundry industry. For the past six years, Ed has provided training, support, and product demonstrations on Autodesk based products, with three of those years working product support for Autodesk's Mechanical Division.

DEDICATIONS

This book is dedicated to my wonderful wife Lisa, and my children, Shawn and Riley. *—Ed O'Halloran*

I dedicate this book to my wife, Cammi, and my children, Allison and Jonathan. Their love helped make this book possible. *—Daniel T. Banach*

General Feature Enhancements

This chapter introduces several new "ease of use" features that have been added to Mechanical Desktop for release 5. These enhancements include: Browser Selection, Feature Color, Transparent 3D Orbit, Lighting Control, New Termination Options, Face Selection, and Design Variable Creation. These new features provide greater maneuverability in Mechanical Desktop.

After completing this chapter, you will be able to:

- Select Multiple Features in the Browser
- Apply Colors to Individual Features
- Invoke the 3D Orbit Transparently
- Control Ambient and Direct Lighting
- New Termination Options for Extrude and Revolve
- Use Face Selection for Placed Features
- Create Design Variables In Process

SELECT MULTIPLE FEATURES IN THE BROWSER

In Mechanical Desktop 5, for certain operations it is now possible to select multiple features in the Browser. Currently, when multiple features are selected in the Browser, the new feature color and feature pattern options are available. To select multiple features, depress the CTRL key and use the left mouse button to select each feature in the Browser. When selected, each feature is highlighted (see Figure 1–1). Once the selection process is complete, right-click to access the options available for the features selected.

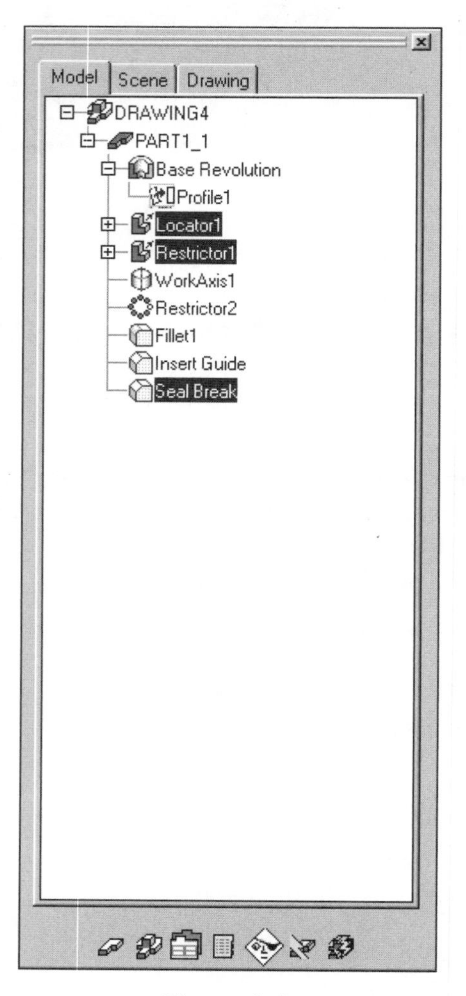

Figure 1–1

APPLY COLOR TO FEATURES

Mechanical Desktop 5 incorporates a new visualization tool, the ability to apply color to individual or multiple features. Applying color to features provides instant visual feedback. Feature Color can be used to group common features, to distinguish between common features of different sizes, or to illustrate mating parts or faces. Figure 1–2 shows how to apply Feature Color, by using the Browser to select the desired feature or features, and right-clicking to access the menu for color selection. The result of applying the same color to two features of a part are shown in Figure 1–3.

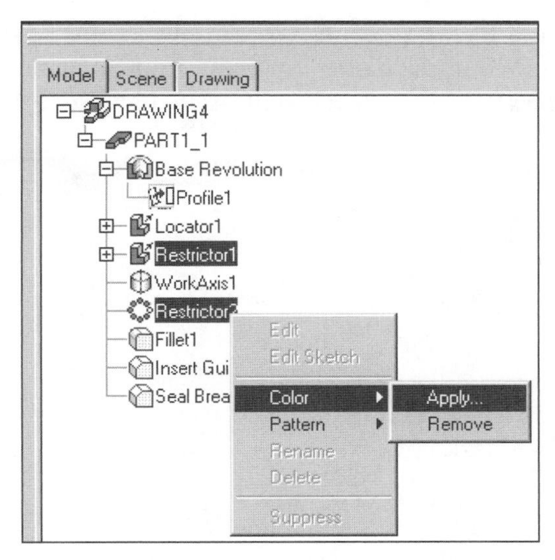

Figure 1–2

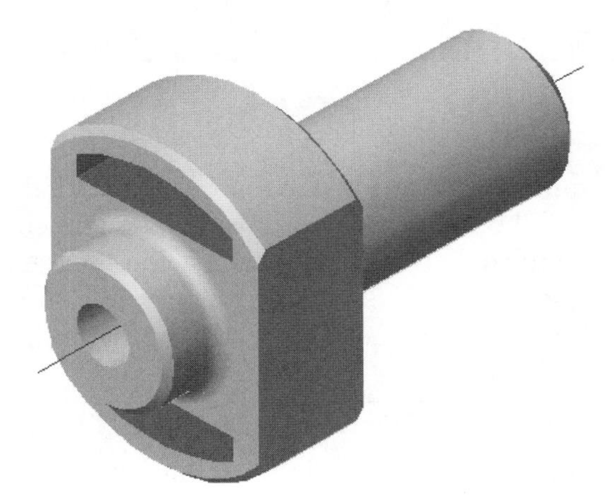

Figure 1–3

TRANSPARENT 3D ORBIT

It is now possible to access 3D ORBIT transparently while executing sketched and placed features commands. Accessing 3D ORBIT transparently allows you to rotate the current view while another command is in progress. This provides improved visibility of feature previews for sketched features, and the ability to select edges or faces anywhere on a part while the command is in progress. The command in progress will determine specific location to access the 3D ORBIT option.

For both Sketched and Placed Features you can invoke the 3D ORBIT command by right-clicking in any field in the dialog box that can use spinner controls for input. An example using the AMHOLE command is illustrated in Figure 1–4.

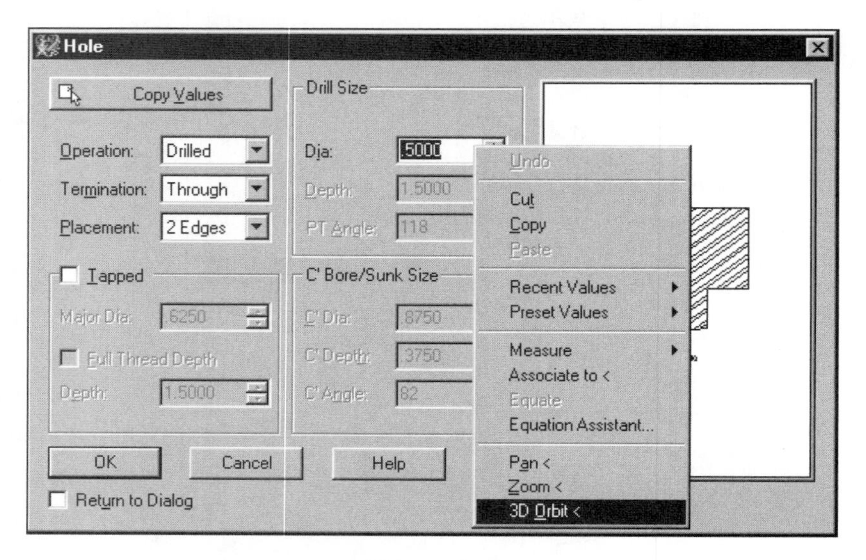

Figure 1–4

Once the 3D ORBIT command has been invoked, the dialog box for the current command will disappear to allow full orbit control, and will remain hidden until the 3D ORBIT command has been terminated. The ability to access the 3D ORBIT transparently can be affected by the display settings. If 3D ORBIT is not available during the execution of sketched and placed features, try reducing the color setting of the system.

CONTROL AMBIENT AND DIRECT LIGHTING

The new AMLIGHT command utilizes three different options that give you control over the display of parts and assemblies while working in a shaded mode. The AMLIGHT command controls both the intensity of Ambient and Direct lighting, and the direction of Direct Lighting. Slider Bars are incorporated for control of Ambient and Direct lighting and a button for accessing the light direction. The AMLIGHT command is accessed from the Mechanical View toolbar (see Figure 1–5), or by typing AMLIGHT at the command line.

Ambient lighting is the general overall lighting of the current display. The Lights dialog box shown in Figure 1–6, provides slider bars for controlling the Ambient and Direct settings, and a button for accessing the LIGHT DIRECTION command. By default, the ambient light setting is approximately 25%, or one quarter of the way across the slider bar starting from the left. Sliding the control to the left will reduce the amount of ambient light, and sliding to the right will increase the ambient light.

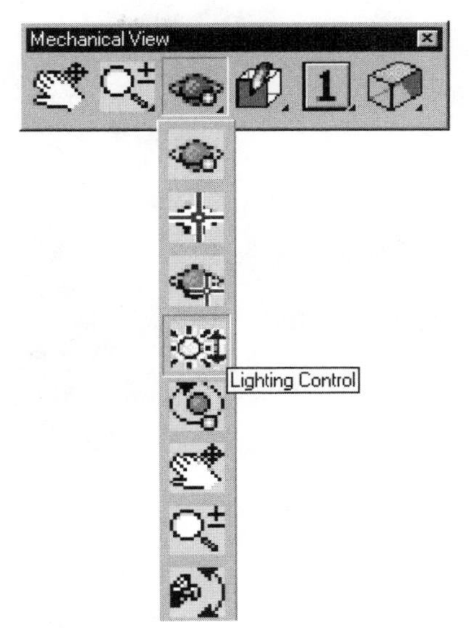

Figure 1–5

Figure 1–6

Direct lighting is the intensity or brightness of lighting used in a shaded mode. As shown in Figure 1–7, the default setting for direct lighting is 100%. This setting can make it difficult to see profiles when viewing along the current sketch plane, or to see part edges. As shown in Figure 1–8, reducing the direct lighting to a setting between 50% and 75% will aid in these areas.

The button just to the right of the slider bar controls sets lighting direction. Once selected, a command line prompt will ask for a point defining the light direction. You can define the direction by selecting in the graphics area (OSNAPS will work here) of the file, or by entering specific coordinates at the command line. As with other viewing options on the Mechanical View toolbar, AMLIGHT is a viewing aid, and is not intended to recognize the location of other parts or features and cast shadows, or to provide high quality renderings.

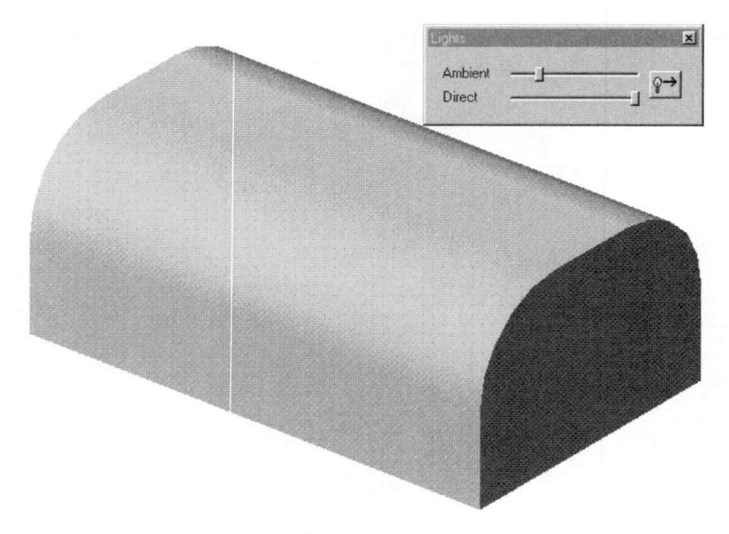

Figure 1–7

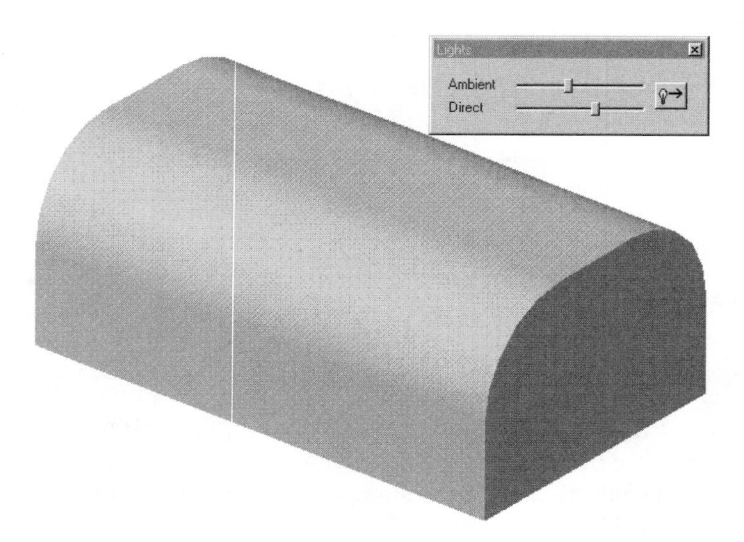

Figure 1–8

The AMLIGHT dialog box is a modeless dialog box and behaves in much the same manner as the Mechanical Desktop Browser and AutoCAD's Object Properties dialog box. It can be docked on either side of the graphics area, or remain floating anywhere on screen, and will remain open until it is closed. The AMLIGHT dialog box settings will reset to default each time shading is toggled.

NEW TERMINATION OPTIONS FOR EXTRUDE, REVOLVE AND SWEEP FEATURES

The termination options for Extrude, Revolve and Sweep features have changed for all features, with the exception of base features. Two new termination options have been added; Next and Extended Face, and the To-Face/Plane option has been split into individual termination options. No termination options have been deleted. Next termination (Figure 1–9) has been added to allow the profile to terminate on the next face that the profile fully intersects. The Extended Face option (Figure 1–10) allows a profile to terminate to a specified face with the profile fully or partially intersecting the selected face.

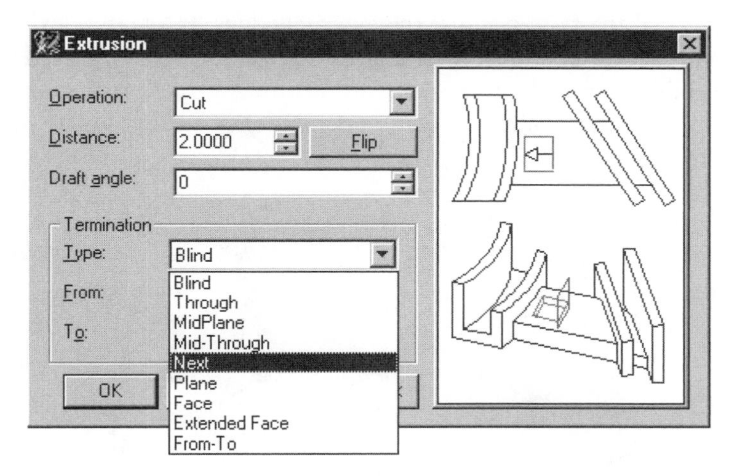

Figure 1–9

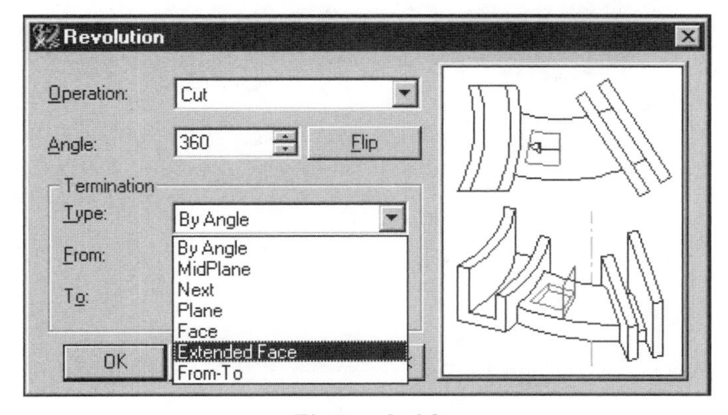

Figure 1–10

FACE SELECTION FOR PLACED FEATURES

Mechanical Desktop 5 now allows selection of all edges of a face in a single pick. In previous releases, to select all edges of a face to add a fillet or chamfer, each edge had to be selected individually. Now, once appropriate information is entered into the dialog box, move the cursor over the face and left pick. As shown in Figures 1–11 and 1–12, Mechanical Desktop will find all the edges of the face and apply the feature. This new functionality will reduce the number of picks needed, and reduce the chance of having to go back and add features, or miss the feature completely.

Face selection cannot be used where fillets or chamfers already exist. Face selection will find all edges of the face, not just the external edges. If a feature exists that does not require the Fillet or Chamfer, then that feature should be suppressed prior to the use of face selection. After the face selection feature is placed, reorder is not an option, the fillet or chamfer will have a feature dependency on that feature that will not allow it to be reordered.

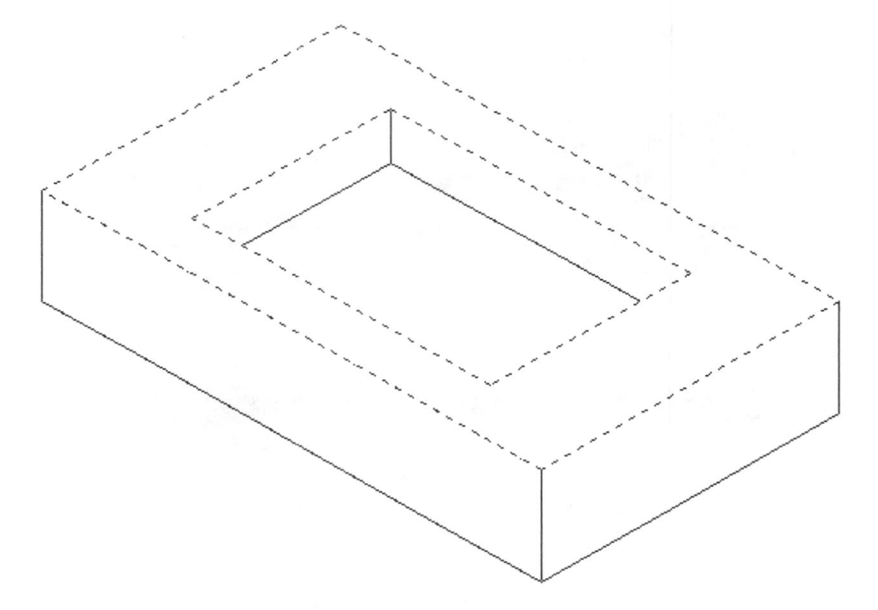

Figure 1–11

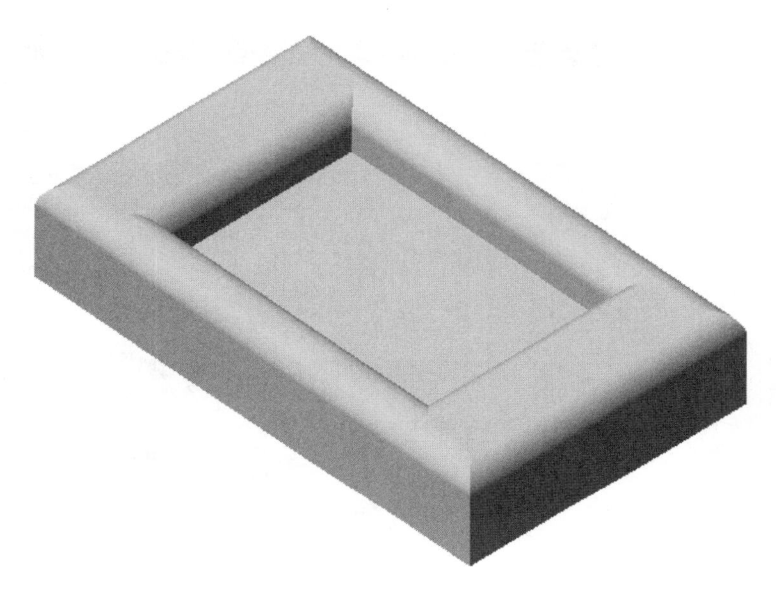

Figure 1–12

CREATING DESIGN VARIABLES IN PROCESS

The ability to build relationships between features on single or multiple parts via Design Variables is a very powerful tool in Mechanical Desktop. Entire assemblies can be updated simply by changing a value in a text file or spreadsheet. Previously, Design Variables had to be created prior to executing a command that would use the variable. Design Variables can now be created and utilized at any time in the modeling process.

To access Design Variables in the process of creating a new feature, access the Equation Assistant and then right-click anywhere in the Active Part Variables or Global Variables fields. As shown is Figure 1–13, context menu will appear with the options to create a new variable, or delete a variable that already exists. To create a new variable, select New, and enter in the values as needed. Once the information is added, double-click on the newly created variable and it will populate the Result field in the Equation Assistant dialog box.

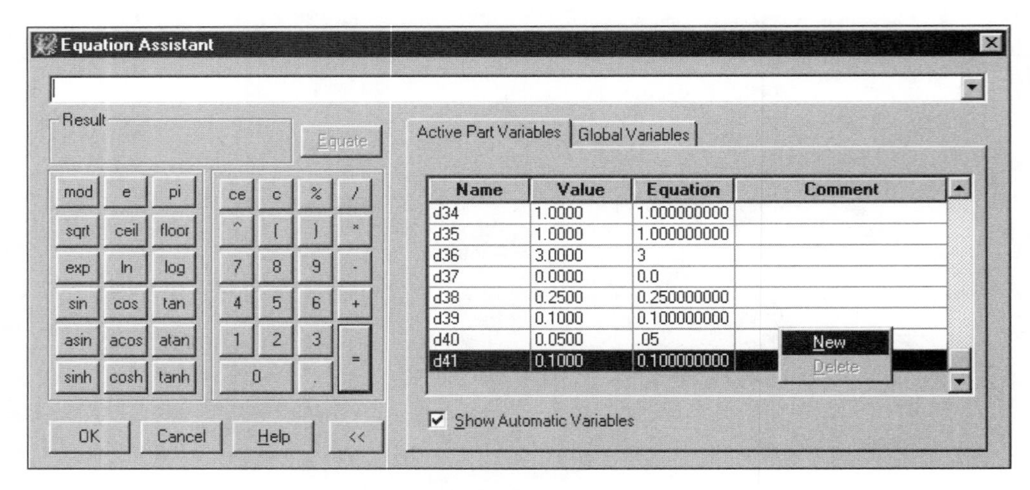

Figure 1–13

Tutorial 1–1 General Feature Enhancements

1. If not already running, start Mechanical Desktop 5.

2. Open file MDO1-EX 1-1.

3. Invoke the AMPARDIM command. Select the circular profile on the flat end of the part and select a location to place the dimension.

   At the Enter Dimension Value Prompt right-click in the graphics area and select the Equation Assistant from the context menu.

   Right-click in the background area for the Active Part Variables and select New from the context menu.

   Enter a variable name of BORE and click in the equation field.

   Enter a value of 1.125

   Click anywhere in the Active Part Variables area and then double-click on the new variable.

   The variable name should appear in the Result field at the top of the Equation Assistant dialog box. (Note: The dimension disappears upon returning to the graphics area because the AM_5 layer is off.)

   Select OK.

4. Select the AMEXTRUDE command, and pick the circle when prompted with Select profile to extrude.

   Make the following selections in the Extrusion dialog box, as shown in Figure 1–14.

   > Operation: Cut
   >
   > Termination Type: Face

   Select OK and when prompted to: Select Face

   Select the through portion of Hole1 as shown in Figure 1–15. Mechanical Desktop will highlight the selected face.

   If your selection resembles Figure 1–15 right-click to accept.. The Hole1 feature will no longer be highlighted and a preview will appear, as shown in Figure 1–16.

   Right-click again to accept the preview option and ExtrusionToFace2 will appear in the Browser.

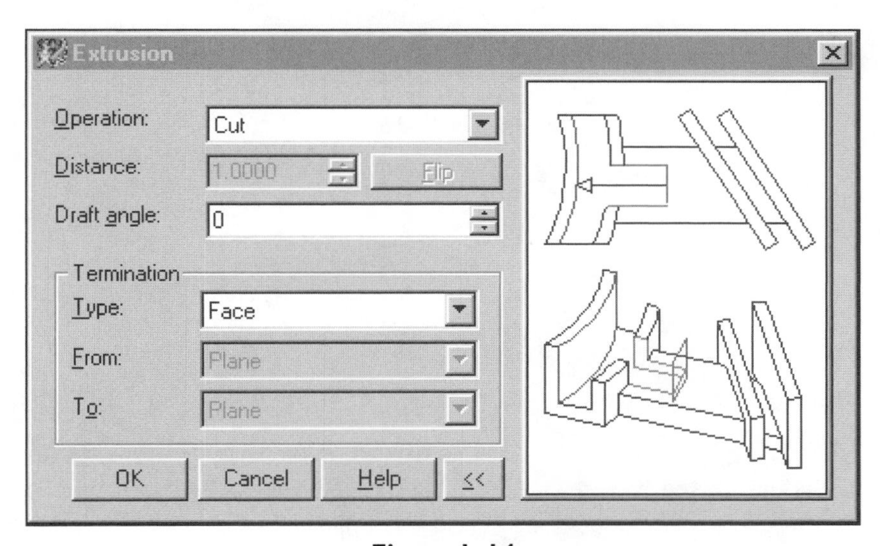

Figure 1–14

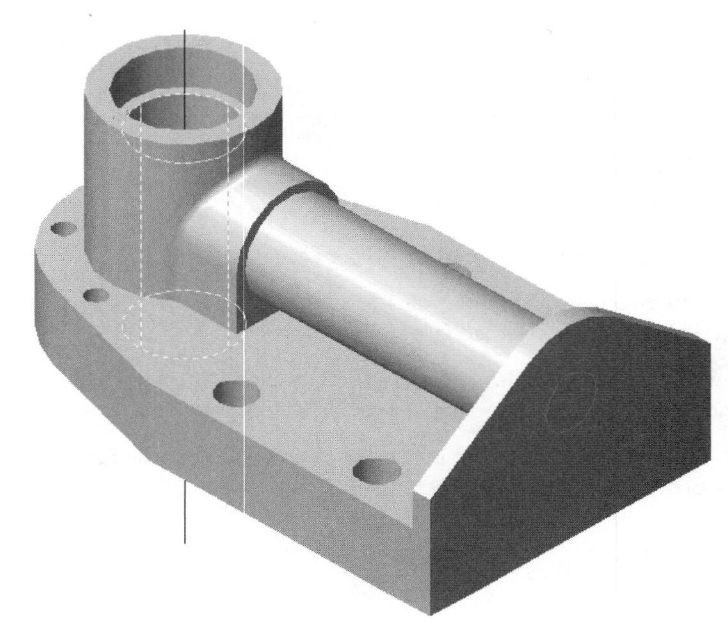

Figure 1–15

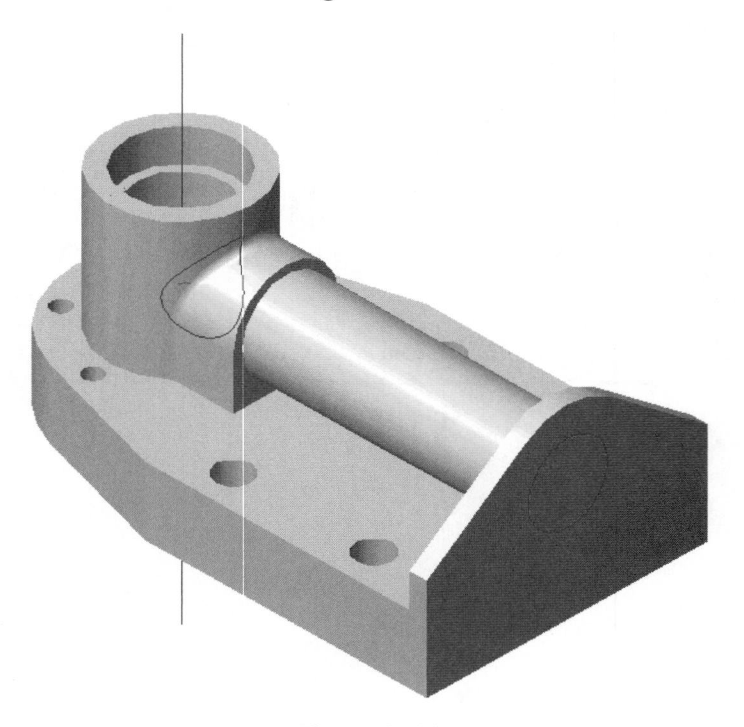

Figure 1–16

5. Right-click on ExtrusionToFace1 in the Browser and select Suppress from the context menu.

   Press the ENTER key (or right-click) to select Yes to continue. Notice that ExtrusionToFace1 disappears in addition to ExtrusionToPlane1.

6. Select UNDO.

7. Right-click on ExtrusionToPlane1 and select Edit from the context menu. The Extrusion dialog box should appear.

8. In the Operation textbox, change the operation from Plane to Next and select OK. The Browser should change from ExtrusionToPlane1 to ExtrusionToNext1 and should be highlighted by a yellow band. This indicates that the feature is in an edited state and requires updating.

9. Right-click in the graphics area and select Update Part.

10. Repeat Step 4 by right-clicking on the ExtrusionToFace1 in the Browser and select Suppress from the context menu. Press ENTER to continue.

    Notice the difference in the part due to the feature suppression from Step 4 to Step 9.

11. Invoke the AMEXTRUDE command.

    Make the following selections in the Extrusion dialog box, as shown in Figure 1–17.

    Operation: Join
    Termination Type: Extended Face

    Select OK and when prompted to: Select Face to Extend.

    Select the outer face of the large cylinder as shown in Figure 1–18.

    Press ENTER to accept the direction.

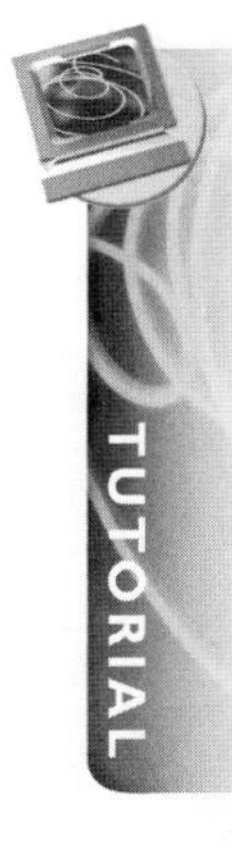

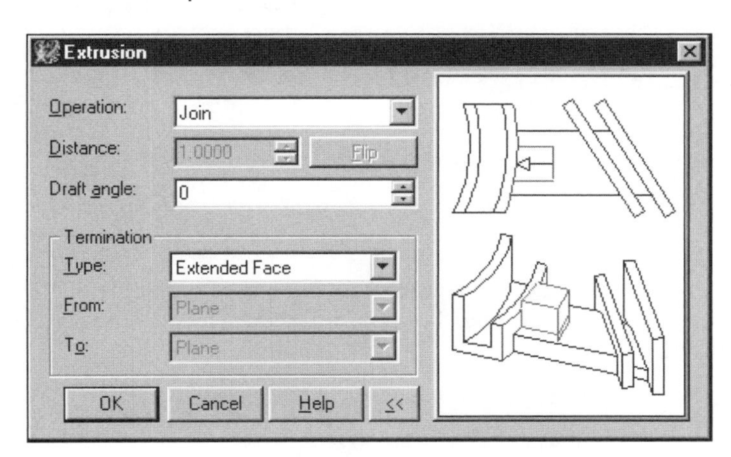

Figure 1–17

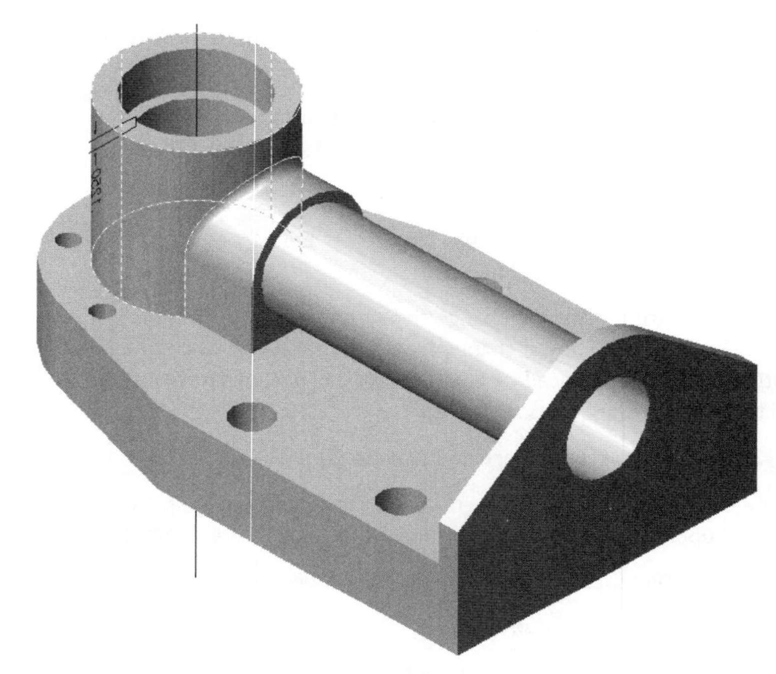

Figure 1–18

TUTORIAL

12. Turn on layer SKETCH1
13. In the Browser, select Hole1 and right-click. Select Color and Apply... from the context menu. When the AutoCAD color index dialog box appears, choose a color of your choice and select OK.
14. In the Browser, hold down the CTRL key and select the following:

    ExtrusionThru1
    Polar Pattern1
    Hole2
    Hole3
    Hole4
    Hole 5

15. Right-click and select Color and Apply... from the context menu. When the AutoCAD Color Index dialog box appears, choose a color of your choice and select OK.
16. If necessary, toggle shading to work in a shaded mode.

17. From the Mechanical View toolbar, select Lighting Control. Slide the controls for Ambient and Direct lighting.

18. Invoke the AMFILLET command and enter a value of .06 for the radius.

At the: Select edges for faces to fillet prompt, right-click in the graphics area and select 3D ORBIT. (Note: You must have "Return to Dialog" checked for this 3D ORBIT mode to work).

Rotate the part until the flat end of the part is in the foreground.

Right-click and select Exit.

Select the end face and hit ENTER twice to apply the fillet to all edges with a single pick.

The completed part should look like Figure 1–19.

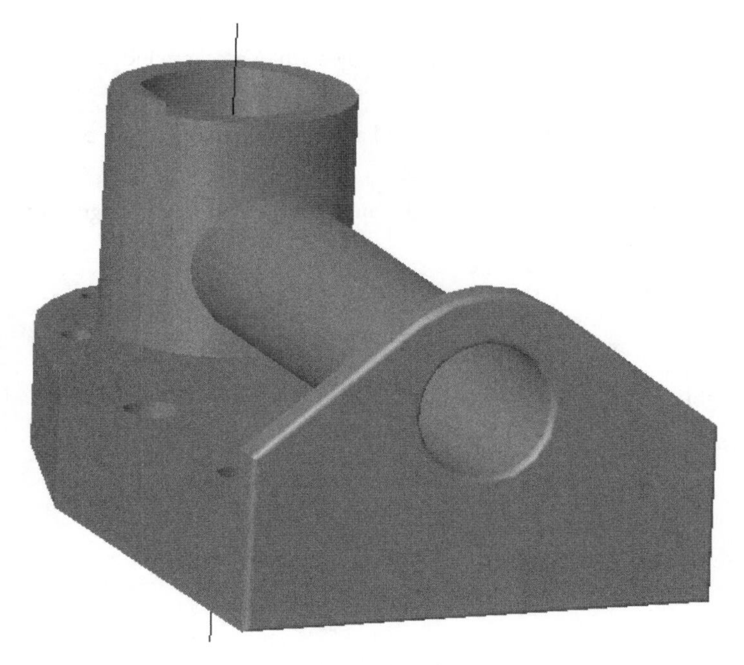

Figure 1–19

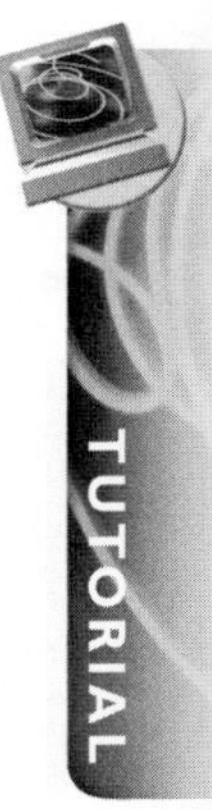

REVIEW QUESTIONS

1. What two items can be applied to multiple features selected in the Browser?
2. What key is held to select multiple features in the Browser?
3. 3D ORBIT can be invoked transparently from the Mechanical View toolbar or by right-clicking. T or F?
4. AMLIGHT settings are retained from drawing to drawing. T or F?
5. Face selection cannot be used where fillets and chamfers already exits. T or F?
6. What is the process to access Design Variables while a command is in progress?
7. Light direction for shading can be changed at any time in the design process. T or F?
8. To exit 3D ORBIT that has been invoked transparently you should: right-click and select Exit, press the ESC key, or either?
9. Once a feature has been assigned a color, it cannot be changed. T or F?
10. Ambient lighting controls the intensity of the light while working in a shaded mode. T or F?

Sketching and Modeling Enhancements

Mechanical Desktop 5 incorporates new sketching tools that allow you to use text to create features and create new open profiles for the new Rib, Bend and Thin Extrusion features.

After completing this chapter, you will be able to:

- Create and Edit Text-Based Sketches and Features
- Create Open Profiles for Rib Features
- Create Open Profiles for Bend Features
- Create Open Profiles for Thin Extrusion Features
- Edit Open Profiles
- Create and Edit Pattern Features

CREATING TEXT-BASED SKETCHES

Mechanical Desktop 5 now allows the use of True Type fonts in sketching. A new command, AMTEXTSK is used to specify and edit the True Type font, style and text to be used. Standard Mechanical Desktop Constraints and Dimensions are used to parametrically locate and size the sketch. These sketches can be used to create base features and to add text to existing parts.

Text-based sketches are created in a similar process to other sketched features. Once the sketch plane has been defined, access the AMTEXTSK command from the Part Modeling toolbar, as shown in Figure 2–1, or from the pulldown menu as shown in Figure 2–2.

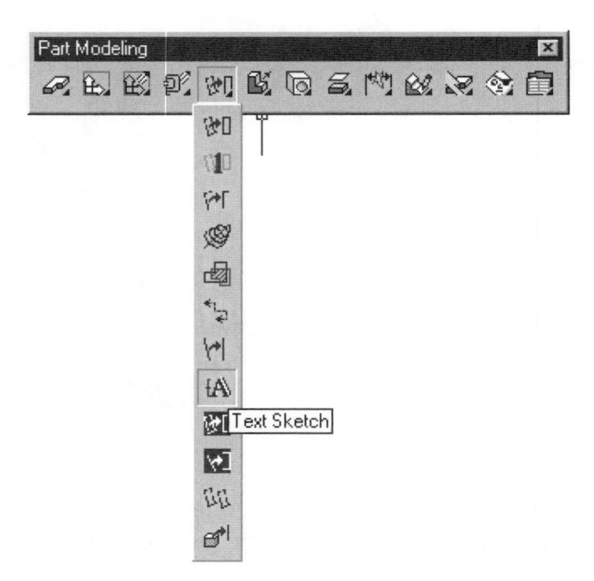

Figure 2–1

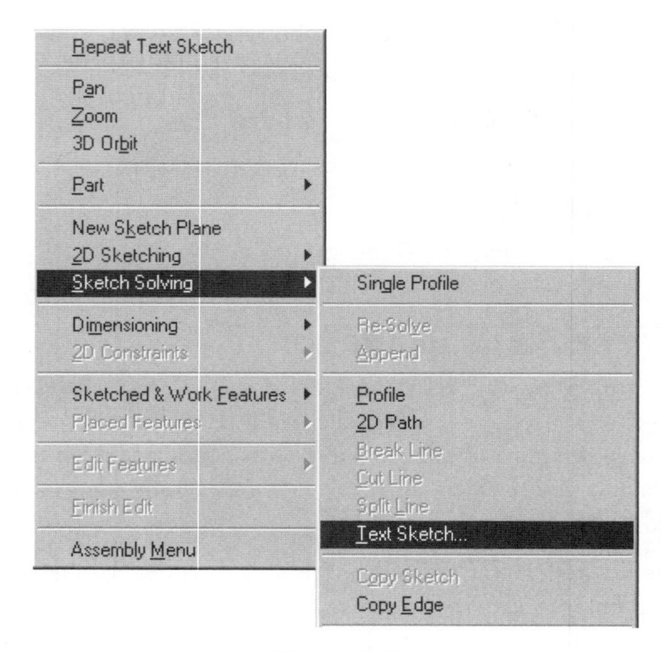

Figure 2–2

In the Text Sketch dialog box, specify the font to be used and the style of the font. In the Text field, enter in the text as you would enter DTEXT in AutoCAD as shown in Figure 2–3. After text has been entered, the OK button becomes active and once selected, you are prompted to specify the first corner of a bounding box on the current sketch plane. A bounding box is used to define the location and height of the text since each font could have any number of vertices to locate it. This bounding box responds to both constraints and parametric dimensions. The text and bounding box are already in a profiled state so there is no need to profile at this point. The dimension that appears defining the height is parametric and can be edited as any other parametric dimension. Add the necessary parametric dimension to fully constrain the bounding box.

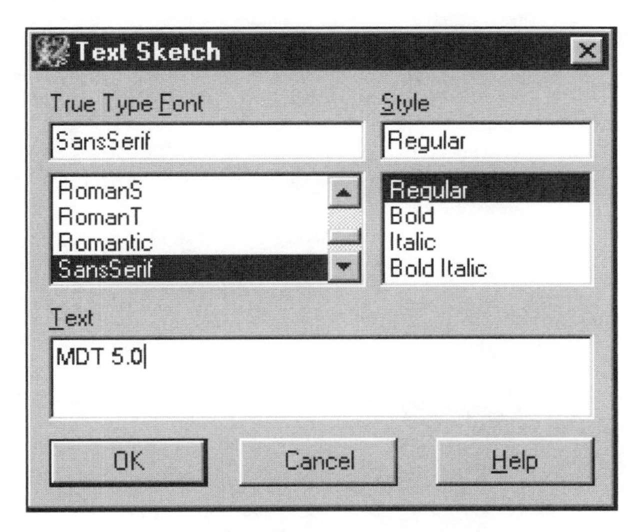

Figure 2–3

CREATING TEXT-BASED FEATURES

Text-based sketches can be turned into Text-Based features using the AMEXTRUDE, AMREVOLVE or AMSWEEP commands provided they follow the normal rules of sketched features. However, only the Blind, Through, MidPlane and Mid-Through termination types are available for the extrusions. By Angle and MidPlane are available for the REVOLVE command. For a sweep, just the Path-Only termination type is available. Draft can be applied to text-based sketches, but due to the usually close proximity of letters to each other, very small draft angles should be used. Figures 2–4 and 2–5 show a text sketch and text feature respectively.

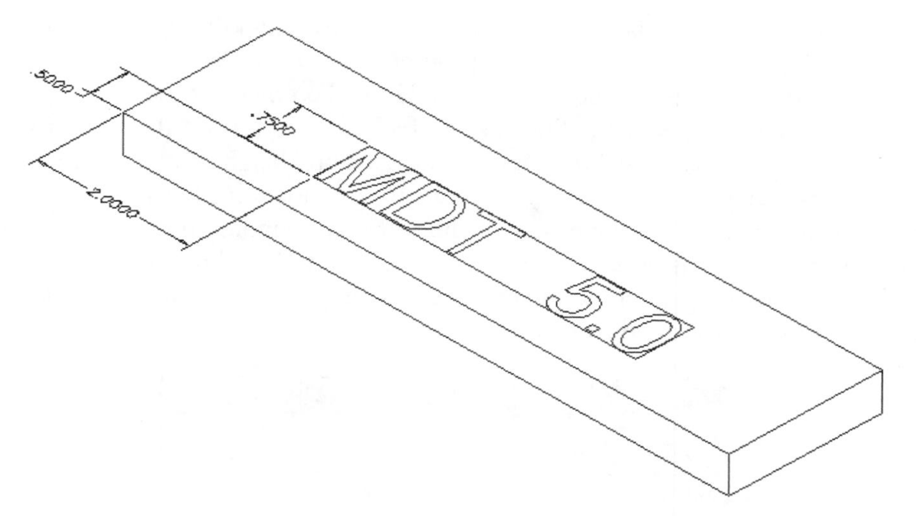

Figure 2–4

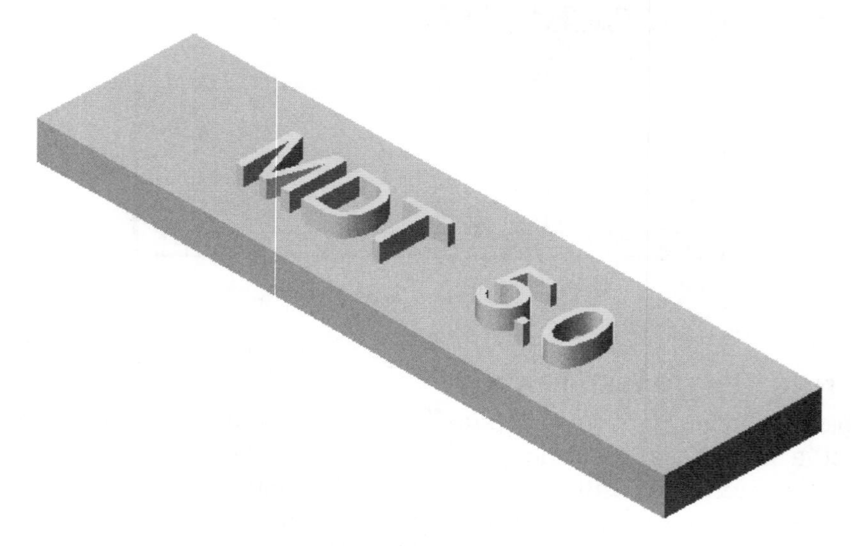

Figure 2–5

OPEN PROFILE FEATURES

Previously in Mechanical Desktop, open profiles were used for creating 2D and 3D paths, cut lines and split lines. Two new sketched features have been added to Mechanical Desktop 5 and one existing command has been enhanced. They are the

Rib (AMRIB), Bend (AMBEND) and Thin Extrusion (AMEXTRUDE) features. Each of these new sketched features requires the use of an Open Profile for their sketches.

An open profile is defined as a profile where the starting point and ending point are not coincident but lie on the same sketch plane. Each feature has a different set of rules that must be applied to us them. Therefore, we will look at each of them individually.

CREATING RIB FEATURES USING OPEN PROFILES

Rib features are used where it would be difficult to sketch and constrain conventional profiles. Sketches for rib features can use any combination of lines, arcs, polylines, splines and construction lines. Once the sketch plane is defined, sketch the shape of the outside shape of the rib. It is not important that the edges of the sketch meet the part, what is important that if extended, the edges of the sketch would intersect the existing part. The reason it is important is that this is what Mechanical Desktop 5 is going to do with the sketch. (see Figure 2–6) Once the sketch is complete, you can profile in a normal manner. Mechanical Desktop 5 prompts you to select an edge to close the profile; however, you should notice a new option at the command prompt. In brackets is the option for open profile as shown in Figure 2–7. If no edges are selected, Mechanical Desktop 5 creates an open profile and displays in the Browser, as shown in Figure 2–8.

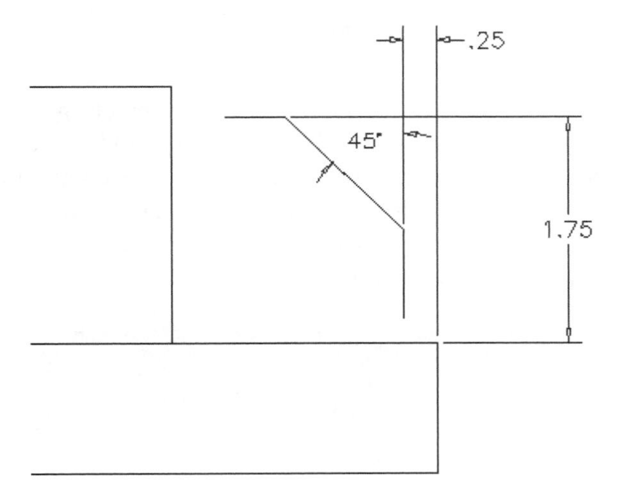

Figure 2–6

```
Select objects for sketch: Specify opposite corner: 3 found
Select objects for sketch:
Select part edge to close the profile <open profile>:
```

Figure 2–7

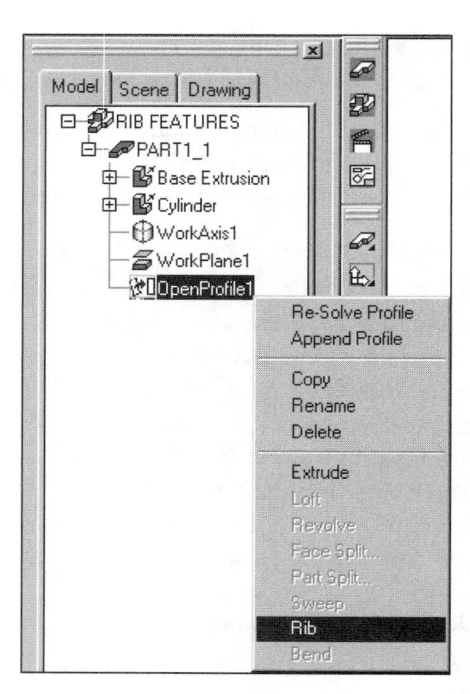

Figure 2–8

After creating an open profile, you have the option of creating a thin extrusion, a bend feature or a rib. Our first example is the rib feature. Rib features are created with the AMRIB command. This command can be accessed via the command line by typing AMRIB, pull-down menus, toolbars or right clicking on the open profile in the Browser. Figure 2–8 shows access from the Browser, Figure 2–9 shows the toolbar button and Figure 2–10 shows access via the menu.

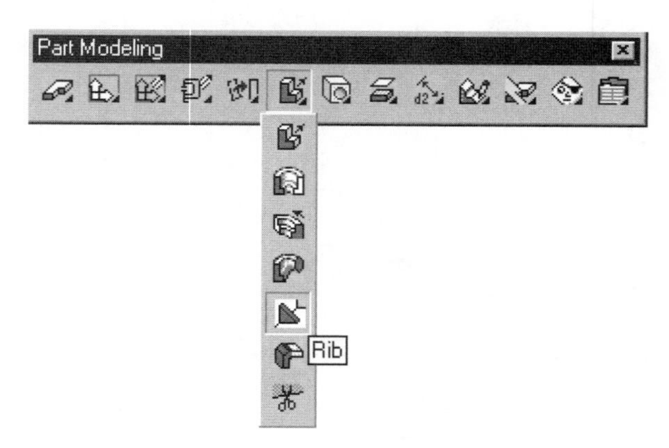

Figure 2–9

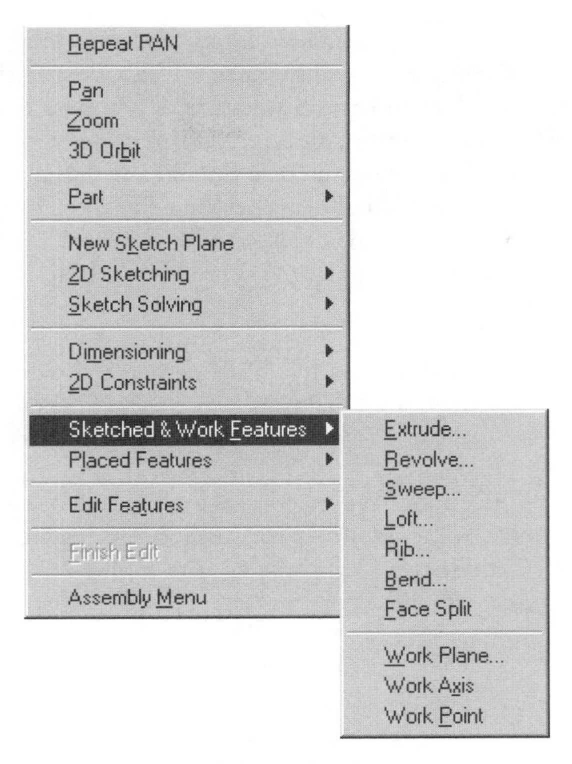

Figure 2–10

The AMRIB command brings up the Rib dialog box and immediately displays a preview that reflects the current settings of the dialog box in the graphics area. In the dialog box, we have four different settings: Type, Thickness, Flip Thickness and Fill Direction, as shown in Figure 2–11.

Figure 2–11

Rib feature types are: One Direction, Two Directions and Midplane. The direction type controls the second feature, Thickness. Thickness, as the name implies, controls the thickness or depth of the rib feature. Thickness is calculated as a Z value, perpendicular to the sketch plane that the open profile was created on. By default, when One Direction is selected, the thickness value is placed in the positive Z direction. This can be changed by checking the third option, Flip Thickness. If the Two Directions type is selected, Thickness 1 is in the positive Z direction, Thickness 2 is in the negative Z direction and the Flip Thickness option is unavailable.

The Midplane option takes the value for Thickness and divides it equally in the positive and negative Z direction. Thickness 2 and Flip Thickness are unavailable with this option.

The last setting is Fill Direction. Fill Direction is represented by the arrow, as shown in Figure 2–12, and must point into the part. Both the endpoints of the sketch and the Fill Direction must terminate into the existing part. Figure 2–13 shows a completed rib feature.

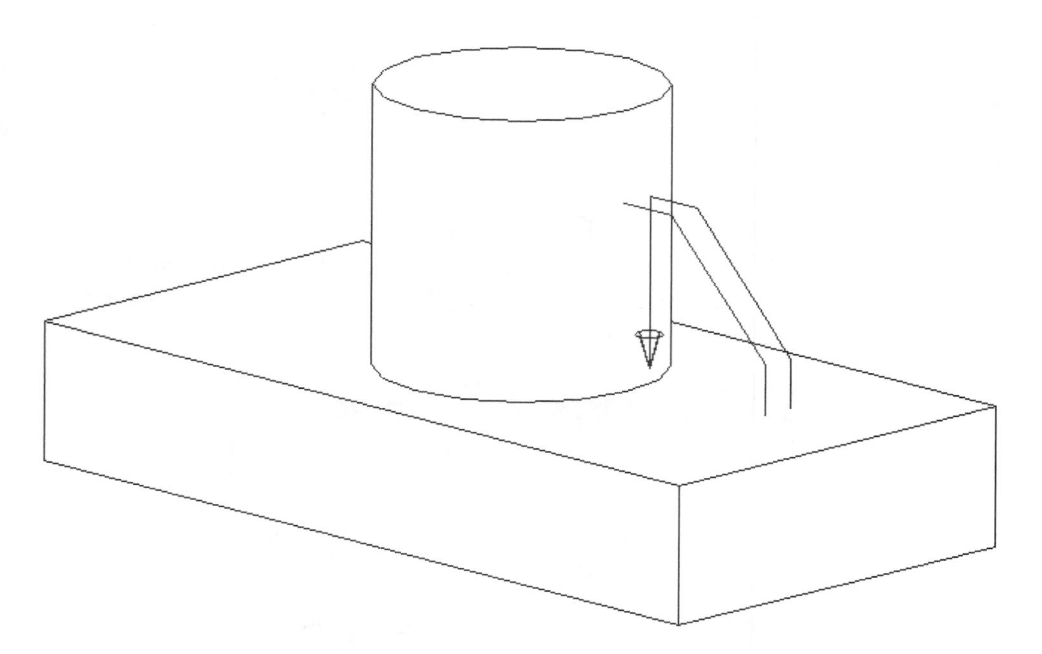

Figure 2–12

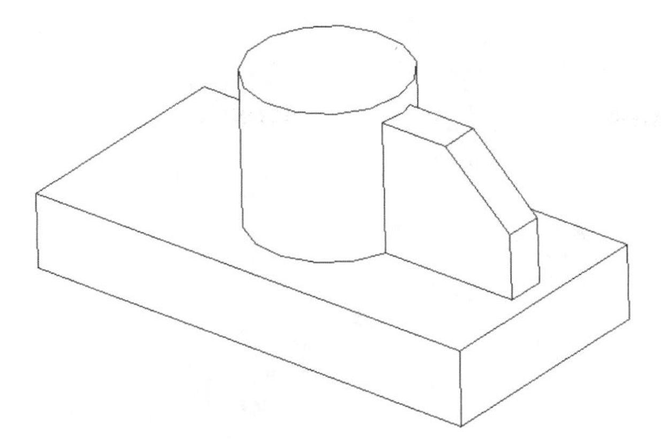

Figure 2–13

CREATING BEND FEATURES USING OPEN PROFILES

Bend features are created based on an open profile consisting of a single line segment. The line segment determines the point of tangency for the bend and must lie on the part. This means that the sketch line determines the start point of the bend, *not* the center of the radius of the bend. The length of the segment can also be a determining factor. As with rib features, profiling a single line segment results in an open profile in the Browser. Figure 2–14 shows an open profile sketch. Notice that the sketch only crosses one section of the part. The AMBEND command can be accessed in the same ways as the AMRIB command. Figure 2–15 shows how you access the command via the Browser.

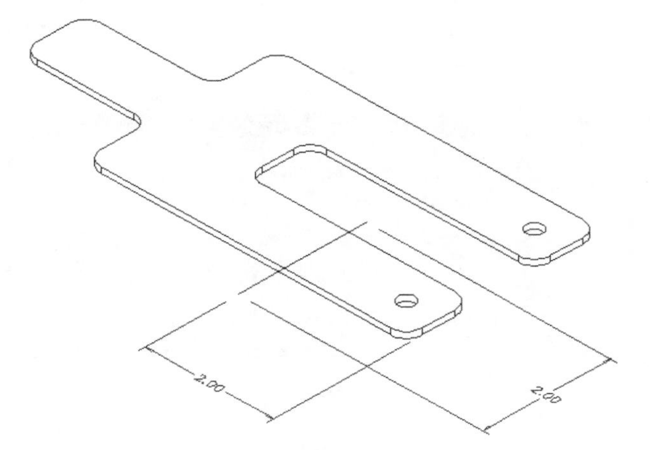

Figure 2–14

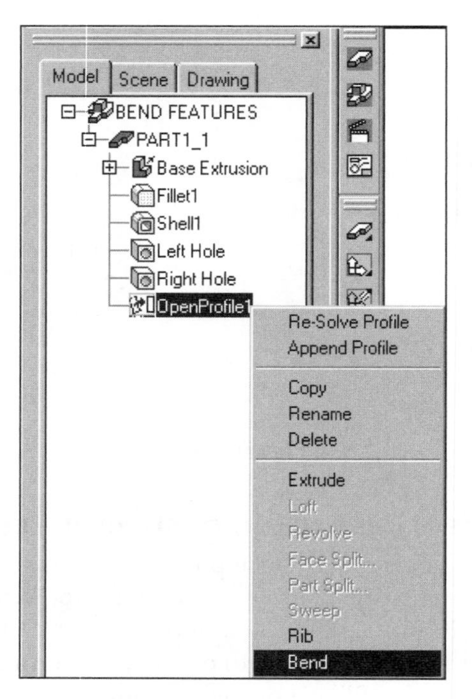

Figure 2–15

The AMBEND command brings up the Bend dialog box, shown in Figure 2–16. The Bend dialog box consists of three fields to input data and two buttons. The three input fields control how the bend is created and the two buttons control the direction of the bend.

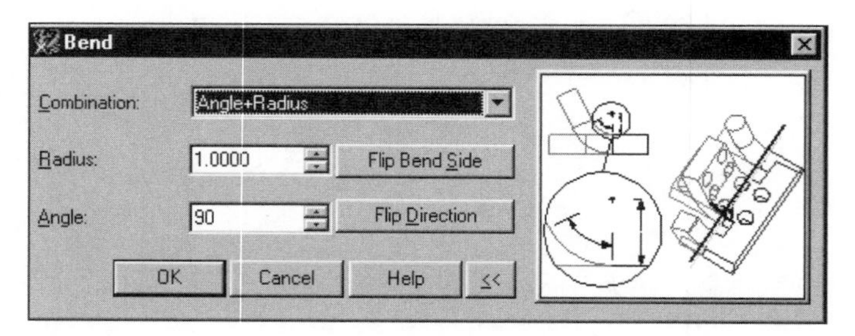

Figure 2–16

The first input field is Combination. In the Combination field you specify three different settings: Angle+Radius, Radius+ArcLen and ArcLen+Angle. The names of each of these combinations clearly define the information needed to complete the bend. With each combination selected, the next two input fields automatically update to reflect the combination selected. Angle+Radius, as the name implies, allows us to specify the number of degrees of rotation and the radius for the bend in the next two input fields. Radius+ArcLen allows a bend about a specific radius and uses the arc length to determine how far the bend progresses. ArcLen+Angle bends a specified number of degrees and derives a radius from the length of the arc. Figure 2–17 displays the Bend Feature Preview and Figure 2–18 displays the completed bend.

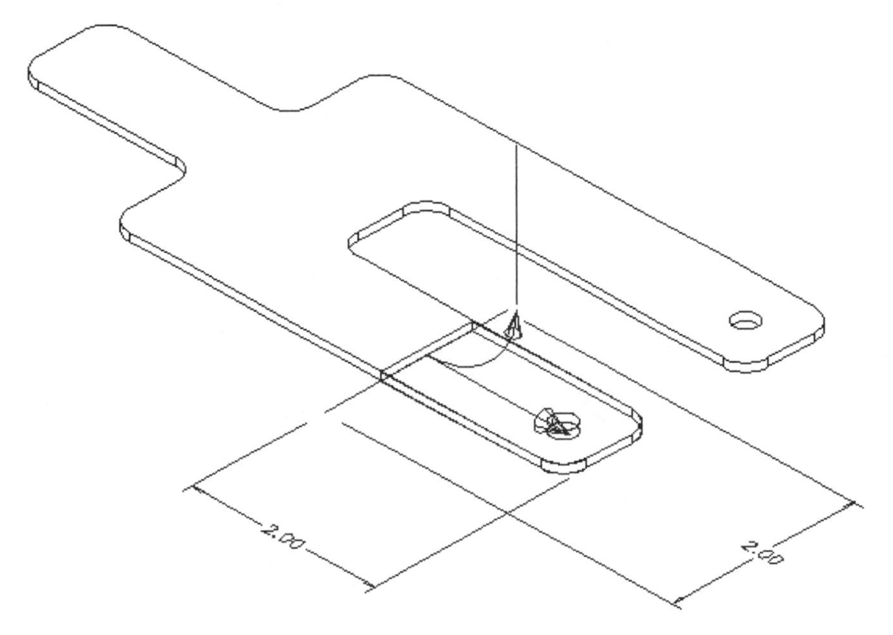

Figure 2–17

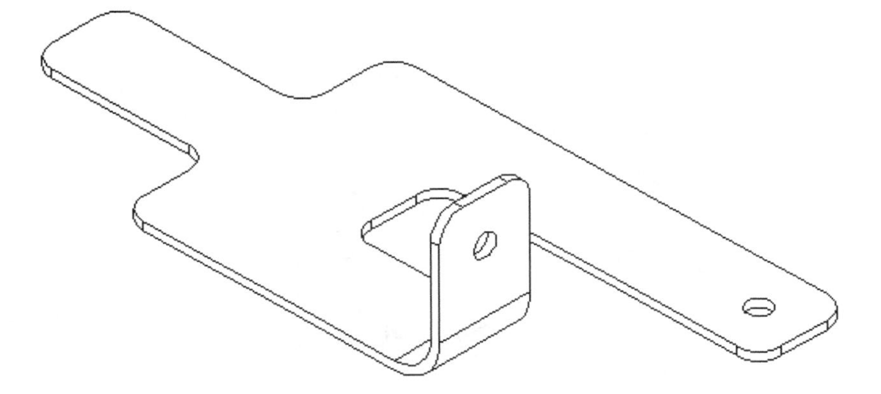

Figure 2–18

Once the proper Combination has been selected, and the appropriate values entered, we use the buttons Flip Bend Side and Flip Direction to fully define the bend. The sketch line divides the part into two halves. The Flip Bend Side button indicates which half of the part bends and displays it in the part preview in the graphics area. The Flip Direction button controls the Z direction. As stated earlier, the length of the open profile for a bend feature might have an effect on the bend. Figure 2–19 displays the previous open profile being edited to increase the length. The result is shown in Figure 2–20. Figure 2–21 displays a bend of a bend.

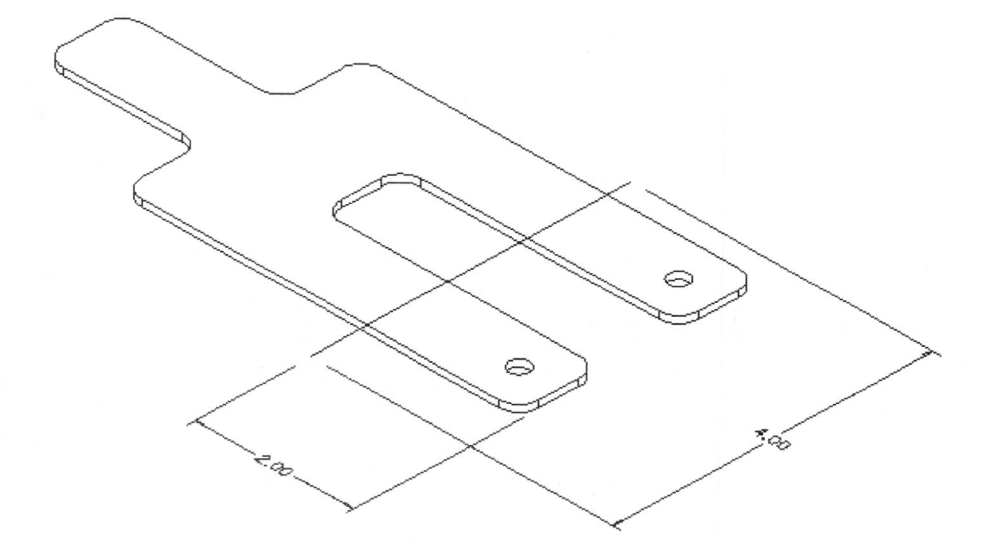

Figure 2–19

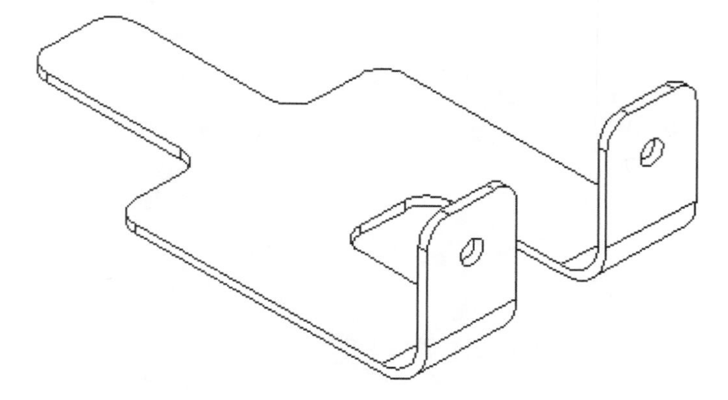

Figure 2–20

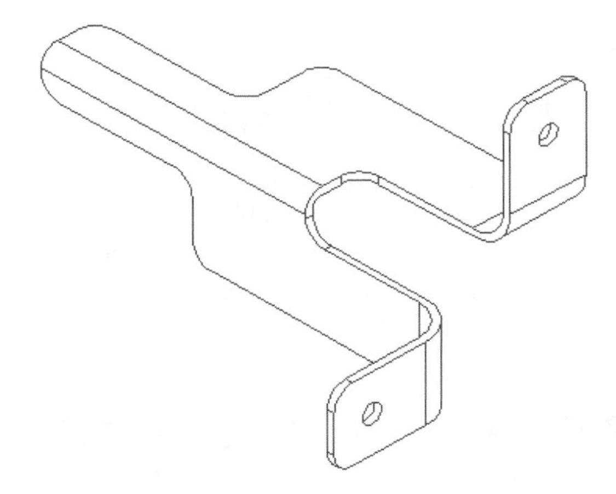

Figure 2–21

While the AMBEND command allows you to bend parts, it cannot be considered a true sheet metal solution. The AMBEND command maintains constant part thickness throughout the bend.

CREATING OPEN PROFILES FOR THIN EXTRUSION FEATURES

Thin extrusion features use any combination of lines, arcs, polylines, splines and construction lines to create an open profile and they use the same AMEXTRUDE command that conventional profiles use. When the AMEXTRUDE command is used in combination with an open profile, additional options are available in the Extrusion dialog box. In Figure 2–22, a single polyline was used to sketch in a support. Notice the sketch goes through an open area in the part and does not extend to meet the outer edges.

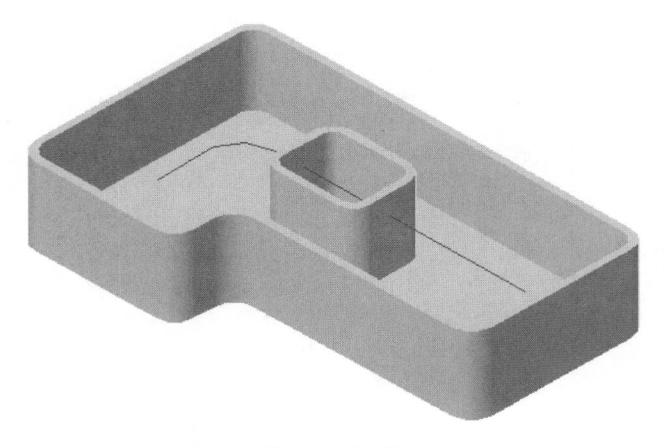

Figure 2–22

With an open profile, a new section called Thickness is added to the Extrusion dialog box as shown in Figure 2–23. In this section, you can set up to five different settings. The first setting is the Type of extrusion to be completed. There are three options for the type: One Direction, Two Directions and MidPlane. The One Direction option allows thickness of the thin extrusion to be defined. This thickness offsets the original open profile on the sketch plane. The Two Directions option offsets the original open profile to each side of the specified distance. The MidPlane option takes the value designated for a single thickness and divides it evenly to offset the open profile to each side. As changes are made in the dialog box, a preview of the changes display in the graphics area. An example of the preview is shown in Figure 2–24.

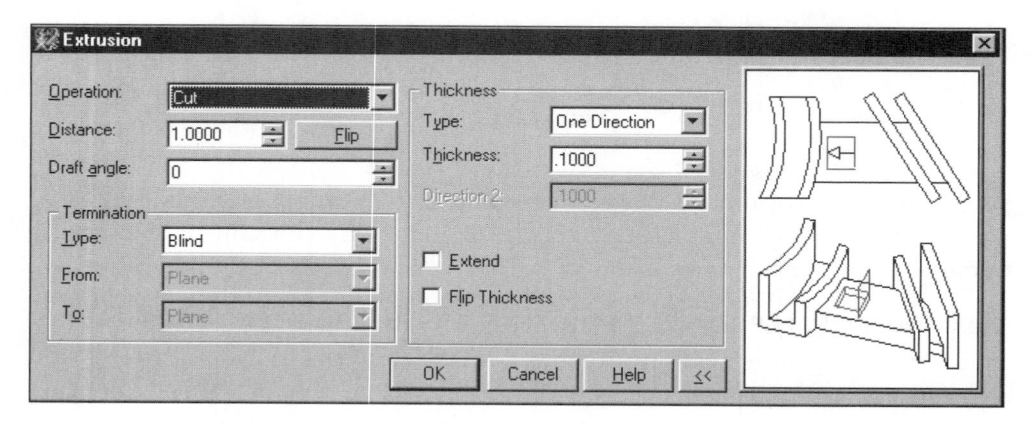

Figure 2–23

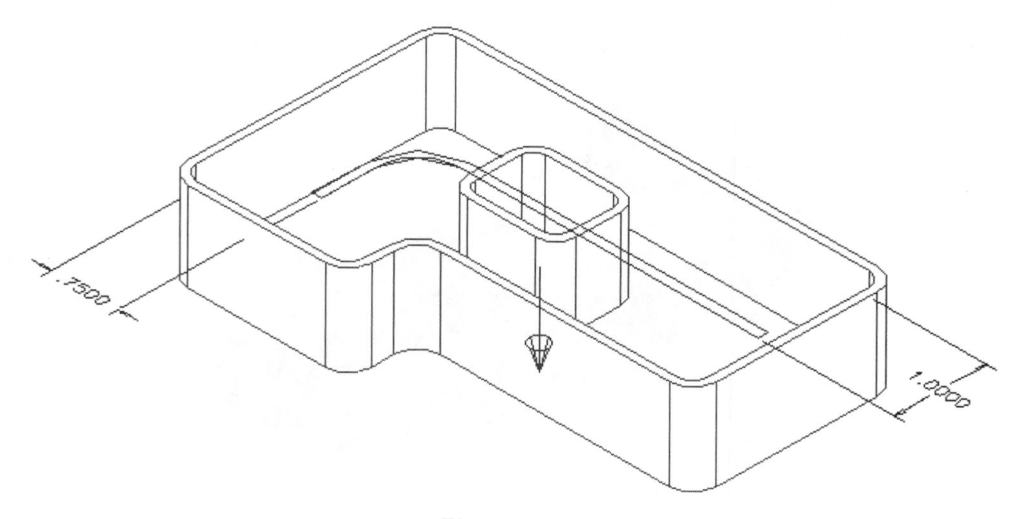

Figure 2–24

Below the controls for thickness, there are two check boxes called Extend and Flip Thickness. The Extend check box is available for all thickness types. Flip Thickness is available only with the One Direction type and changes the offset direction. The Extend option continues the open profile until it terminates into the existing part. If extending the open profile does not intersect the existing part, then that segment of the thin extrusion is not created. Figure 2–25 displays the thin extrusion completed, without the Extend option, while Figure 2–26 displays the result with the Extend option selected.

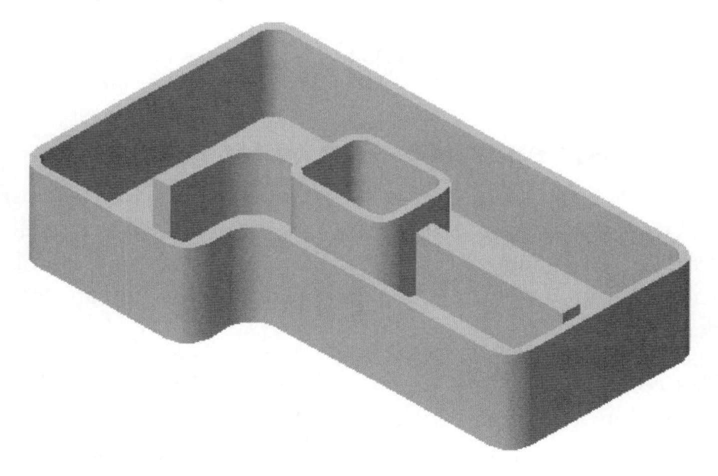

Figure 2–25

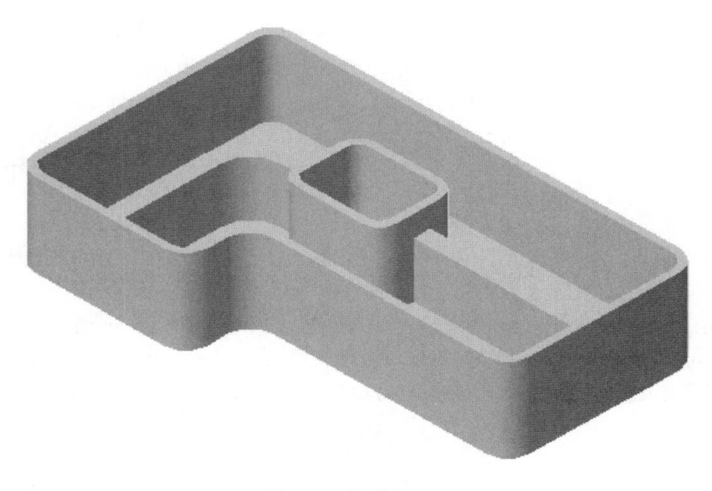

Figure 2–26

Once the process is complete, it displays in the Browser as Thin Extrusion and displays the same icon as a regular extrusion.

EDITING OPEN PROFILES

Editing open profile sketches and features is identical to editing any other sketched feature. You can access the AMEDITFEAT and pick the part in the graphics area or right-click on the feature in the Browser. With either option you can select to edit the feature itself or the open profile that defines the sketch. When selecting the feature, the same dialog boxes that display to create the feature are made available to edit the feature. During the edit process, the previews also display in the graphics area to reflect the changes being made in the dialog box.

CREATING AND EDITING PATTERN FEATURES

The new AMPATTERN command has been developed to replace the existing AMARRAY command. With this new command comes new functionality as well. We can use AMPATTERN to create rectangular and polar patterns just as we have before but with more control. In addition to rectangular and polar patterns, we can also create axial or helical patterns.

The AMARRAY command has not totally disappeared. In the case of migrated files, (files from a previous release of Mechanical Desktop) the AMARRAY command is still used for editing purposes and is only accessible when used in combination with the AMEDITFEAT command. It is not available to create any new features.

AMPATTERN can be used to create patterns of sketched features, extrusions, revolve features, swept features, holes and even combined features. This means that we can create parametric patterns of features on a part and also of the parts themselves. We can also edit existing patterns to add new features and we can now create patterns of already patterned features. Last, we can suppress instances of patterns when they are created or later in the design process. As with the other new modeling features, the AMPATTERN command gives us an in-depth preview of the pattern to be created.

The AMPATTERN command displays the Pattern dialog box, which can be accessed via pull-down and right click menus, Toolbars and the command line. Each of the pattern options; rectangular, polar or axial, has its own button or menu selection; however once in the Pattern dialog box, each of the three options can be accessed. (see Figure 2–27 and Figure 2–28) Like AMARRAY, AMPATTERN is a placed feature.

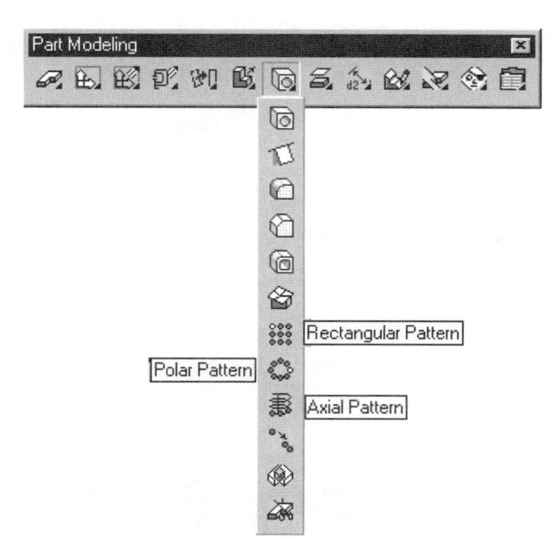

Figure 2–27

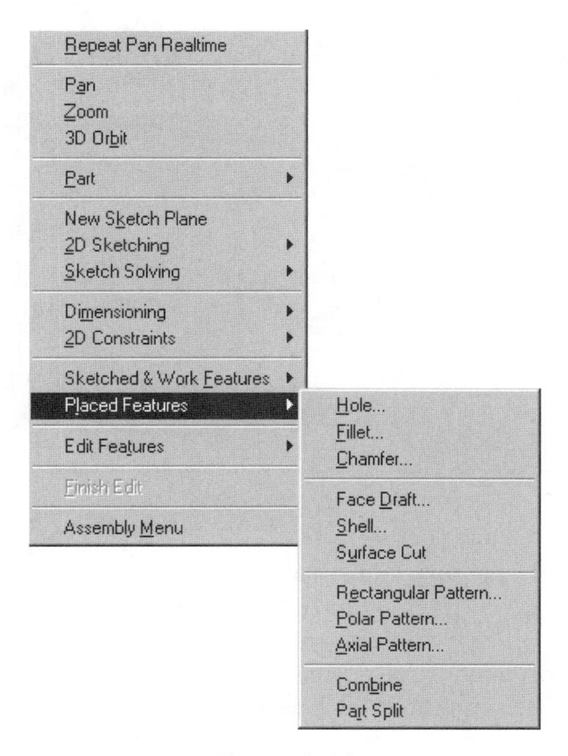

Figure 2–28

The Pattern dialog box contains two tabs: Pattern Control and Features. The Pattern Control tab is where the pattern is built and the Features tab controls the features that are included in the pattern and where new features can be added or existing features removed from the pattern.

The Pattern Control tab is divided into different sections. The first section allows you to choose a Type (Rectangular, Polar or Axial), and includes buttons to Suppress Instances and to define the Plane Orientation. The next sections control Column Placement and Row Placement for Rectangular Patterns, Polar placement for Polar patterns and Axial placement for Axial patterns, respectively. Along the bottom of the dialog box is a button to control preview display and OK, Cancel and Help buttons that are common to both the Pattern Control Tab and Features Tab. Figure 2–29 shows the new Pattern dialog box.

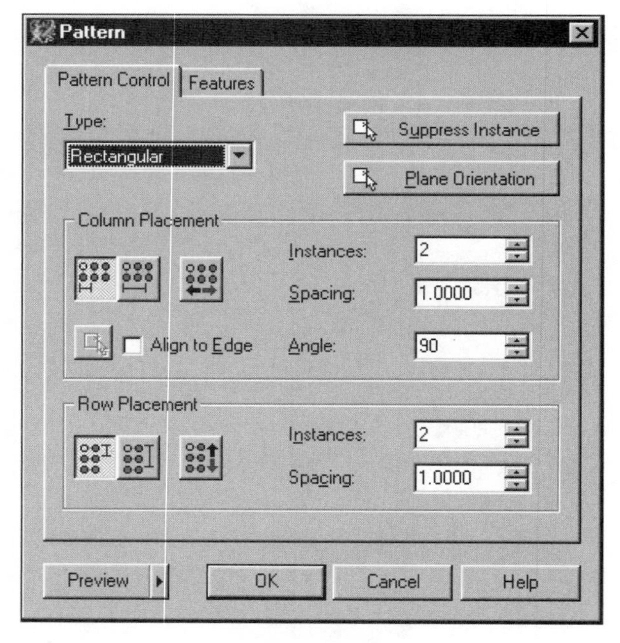

Figure 2–29

Two pattern types, Rectangular and Polar, are common to all releases of Mechanical Desktop. However, in Mechanical Desktop 5, there is more control over the creation of these types. Additionally, the Axial type is new. Regardless of the way the Pattern dialog box is accessed, at any time during the creation of the Pattern, we can change the type of pattern being created. However, once the pattern has been created, it cannot be changed from one type to another.

Rectangular patterns, by default, are created on the current sketch plane. However, the Plane Orientation button allows you to change the sketch plane from within the Pattern dialog box. To do so, click the Plane Orientation button and then click an existing plane on the part.

With a rectangular pattern, we need to define the column placement. We have three settings to define the columns. The first setting is Incremental Spacing. Incremental Spacing defines the center-to-center distance between each column. For example, a one-inch square extrusion with column spacing of 1.5 inches would have half an inch gap between each column. See Figure 2–30 and 2–31 for a comparison between incremental and included spacing. As with the other new part modeling features of Mechanical Desktop 5, the AMPATTERN command gives a full preview of the feature to be created.

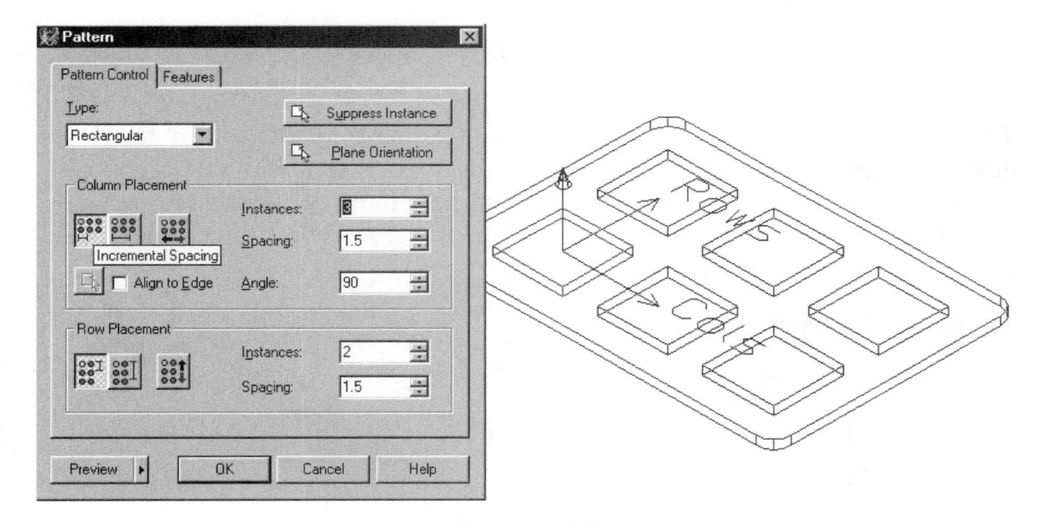

Figure 2–30

The second setting is Included Spacing. Included Spacing defines the total distance the pattern occupies. Taking the example given above, using Included Spacing would mean that the center-to-center distance between the first and last instance in the pattern would be 1.5 inches, regardless of the number of instances in the pattern.

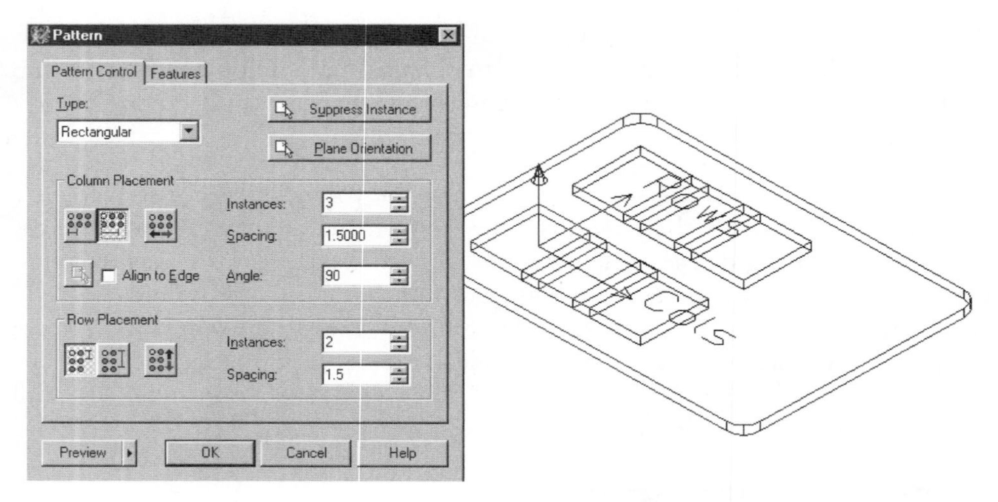

Figure 2–31

The last setting is Flip Column Direction. The setting is used to control the X-axis direction for the column placement. In previous releases we would have had to enter in a negative X value to change the direction of the pattern. As with all settings in the Pattern dialog box, the preview updates immediately to reflect the changes made. Figure 2–32 shows the Flip Column button.

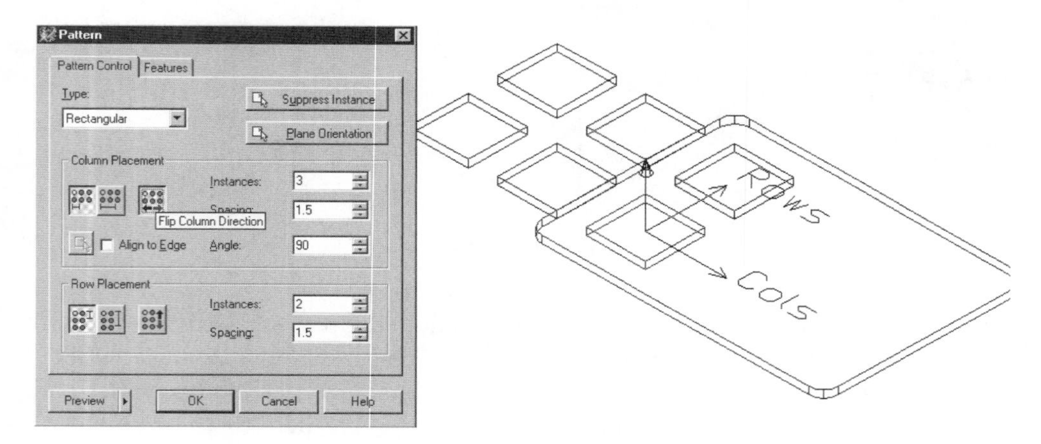

Figure 2–32

To the right of the spacing buttons are three fields to enter in data. The first two, Instance and Spacing, are items common to all releases of Mechanical Desktop. However, the last field, Angle, is new to Mechanical Desktop 5. With this setting, you can parametrically control the angle at which the pattern is created. In previous releases, we were restricted to creating the pattern at the 90-degree angle of the XY plane. In addition to specifying an angle, we can also use the existing part to parametrically define the angle of the pattern. Just below the buttons for controlling the spacing is a check box called Align to Edge. Placing a check in the box grays out the angle field and activates the button just to the left of the check box. Clicking this button turns off the Pattern dialog box, returns you to the graphics area, and prompts you to select an edge or work axis for column alignment. Using this option would mean that if the angle of the selected edge or axes were to change, the pattern would update as well. At any time, the pattern can be edited from a specified angle to align to edge and vice versa. Figure 2–33 displays a preview of the angle option, while Figure 2–34 displays the result when aligning to an edge of a part.

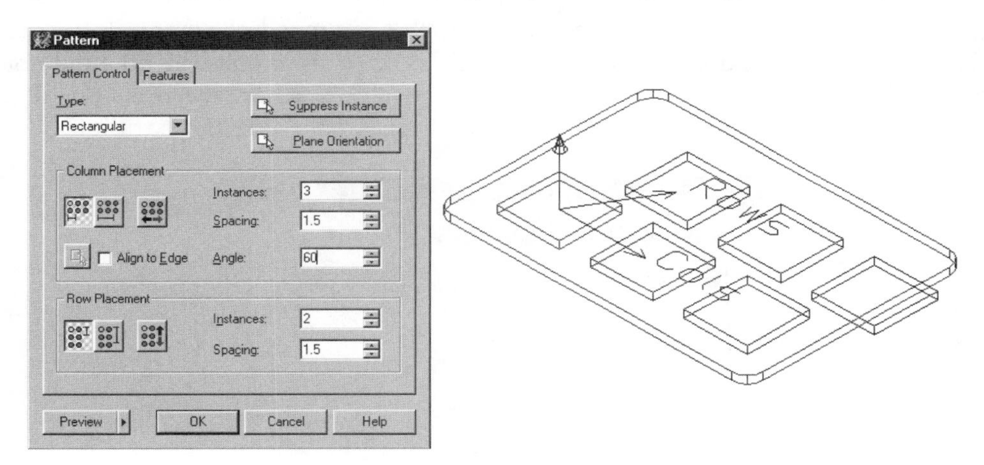

Figure 2–33

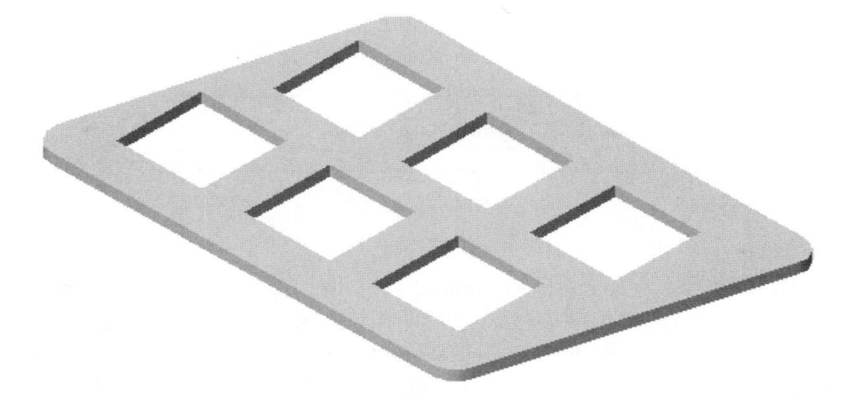

Figure 2–34

The last section of the Pattern Control tab is Row Placement. The settings in Row Placement are identical to those in Column placement with the exception of the Angle field and the Align to Edge check box are omitted. The first button controls the Incremental or center-to-center distance of the rows and the second button controls the total distance the rows of the pattern occupy.

The Features tab shows a list of all the features included in a specific pattern as shown in Figure 2–35. At any time, features can be added to or deleted from the pattern. However, feature order and feature dependencies have to be considered. Taking the previous example of the 1-inch square extrusion pattern, if we added fillets to the inside of the initial extrusion after the pattern feature had already been created and wanted those fillets to show on every instance of the pattern, we could not just edit the pattern and add them in. During an AMEDITFEAT command, the part rolls back to the state when the feature being edited was created. At this point, the fillet did not exist and would not be available for selection. We would first have to reorder the fillet to display in the browser before the pattern was created but after the extrusion.

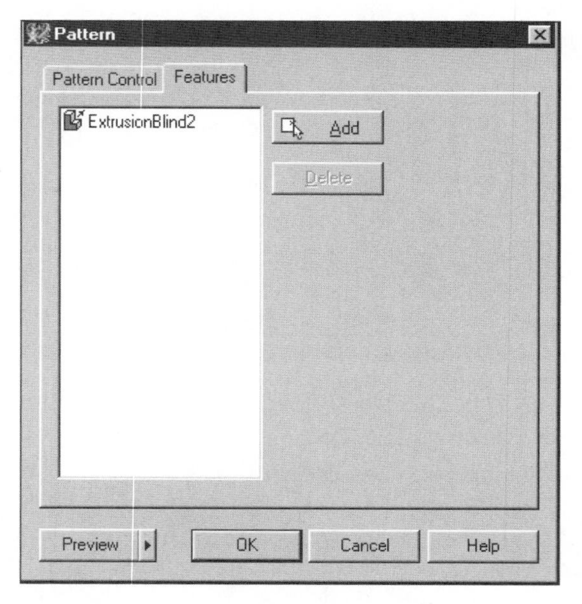

Figure 2–35

At the bottom of the Pattern dialog box is a flyout button for the Preview, as shown in Figure 2–36. If the arrow to the right of the button is selected, you have two options: True Preview and Dynamic Preview. If True Preview is checked, the preview displays an exact representation of the settings defined by Column and Row placement. If unchecked, the preview is a basic representation of the pattern. If Dynamic Preview is checked, the preview updates immediately as settings in the dialog box are changed. If unchecked, the Preview button has to be selected to display any changes made in the Pattern dialog box.

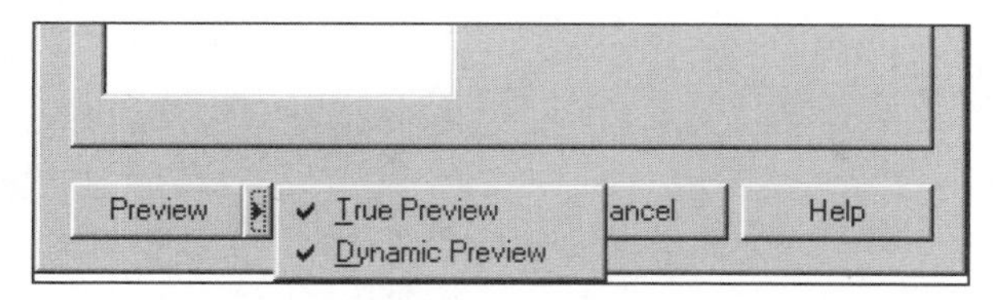

Figure 2–36

Tutorial 2–1 Text Sketch, Text Feature, Open Profiles and Rib Features

1. Start Mechanical Desktop 5 if it is not already running.
2. Open file MD02–EX 1.dwg; the file should look like Figure 2–37

   First we are going to place text over the cutouts and then edit the font style of that text.

   Then we will add a rib feature down the center of the part and edit that sketch to change the end result.

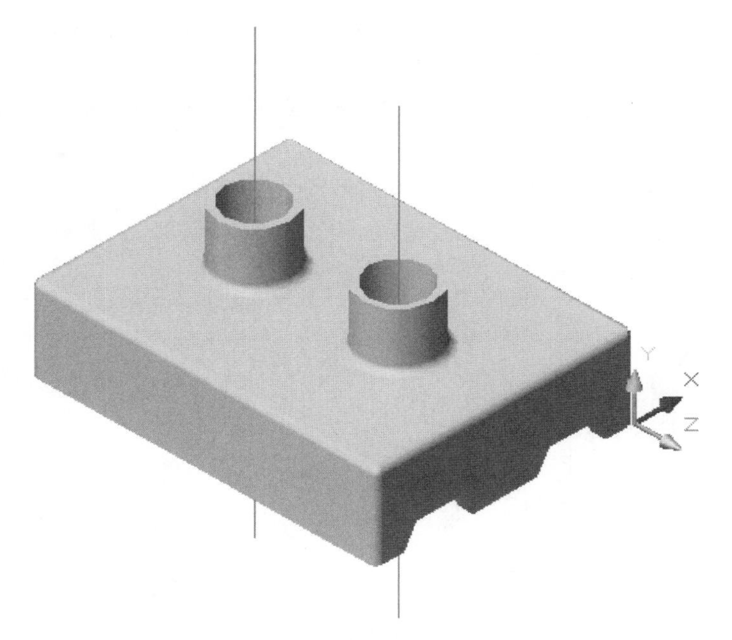

Figure 2–37

3. Place a sketch plane on the front face with the two cutouts.
4. Press the 9 key and press ENTER to look directly on the sketch view, zoom in until your screen resembles Figure 2–38.

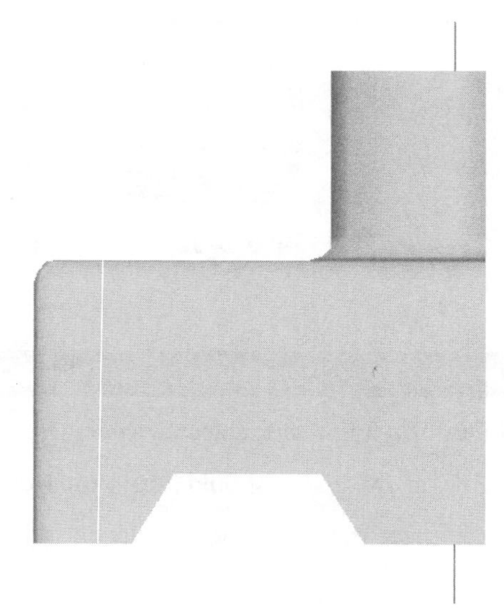

Figure 2–38

5. Right mouse click in the graphics window and from the pop up menu select Sketch Solving > Text Sketch.
6. Type "LIFT" in all caps into the lower field in the dialog box. Use the Tahoma font, regular style.
7. Dimension the sketch as indicated in Figure 2–39.

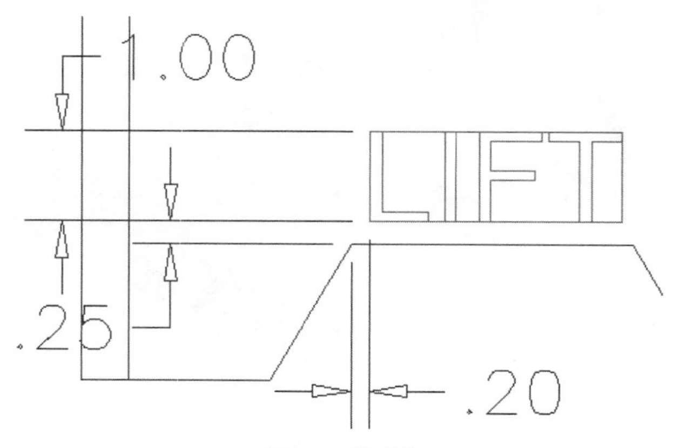

Figure 2–39

8. Extrude the sketch, entering the settings shown in Figure 2–40 into the Extrusion dialog box. Once completed, your part should resemble Figure 2–41.

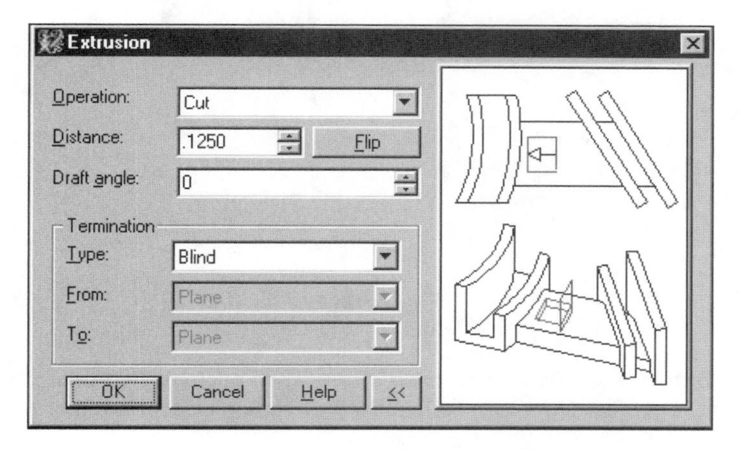

Figure 2–40

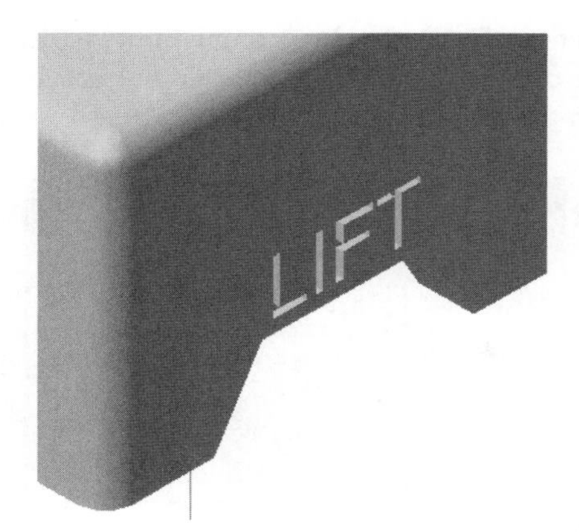

Figure 2–41

9. Press the 8 key to restore an isometric view.
10. Right click on WorkPlane1 and select visibility to view the work plane in the graphics area.
11. Place a sketch plane on the work plane.
12. Press the 9 key and press ENTER to look directly at the sketch view.
13. Using the LINE command, sketch an open profile that resembles Figure 2–42. Be sure that the ends of the sketch will *not* intersect the part if extended.

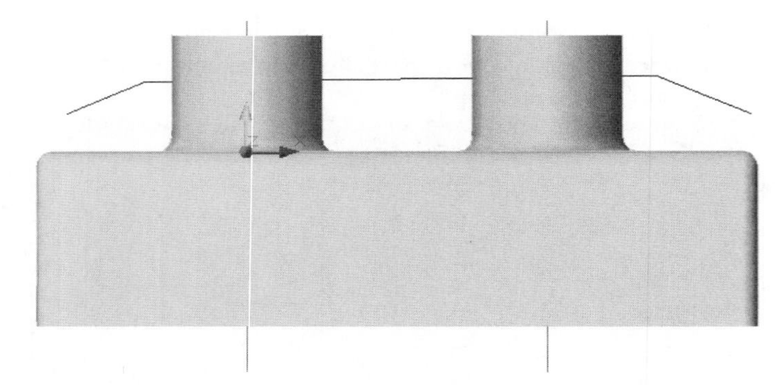

Figure 2–42

14. Profile the sketch. Be sure to press ENTER when prompted to select an edge to close the profile. Once done, OpenProfile1 should display in the browser.

15. Press the 8 key to restore an isometric view.

16. Right click on OpenProfile1 in the Browser and select Rib.

17. Use the settings from Figure 2–43 to complete the RIB command. Be sure that the arrow in the preview is pointing towards the part. Notice the rib is only located between the two ports.

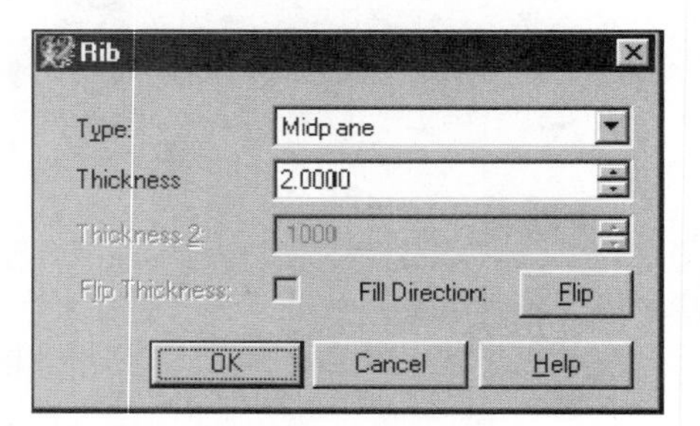

Figure 2–43

18. Press the 9 key and press ENTER to look directly on the sketch view.

19. Right click on Rib1 in the Browser and select Edit Sketch.

20. Grip-edit the ends of the profile so that if extended, the line would intersect the part.

21. Clear the grips and update the part. The rib should now go completely across the part.

22. Press the 8 key to restore an isometric view. The part should resemble Figure 2–44.

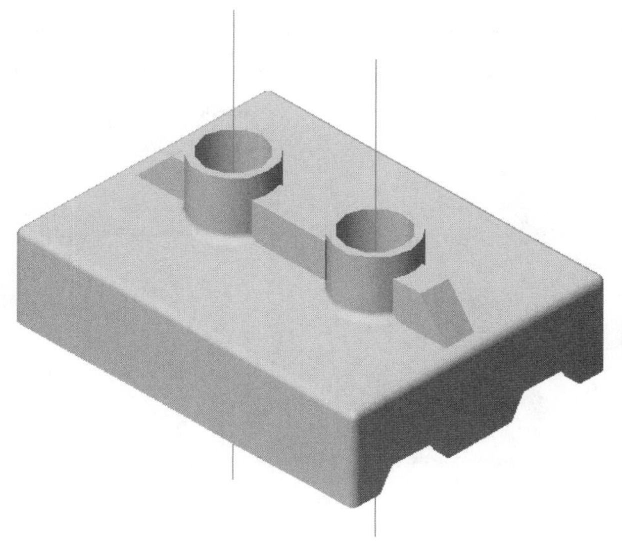

Figure 2–44

23. Save the file.

Tutorial 2–2 Open Profiles and Bend Features

1. Start Mechanical Desktop 5 if it is not already running.

2. Open file MD02–EX 2.dwg.

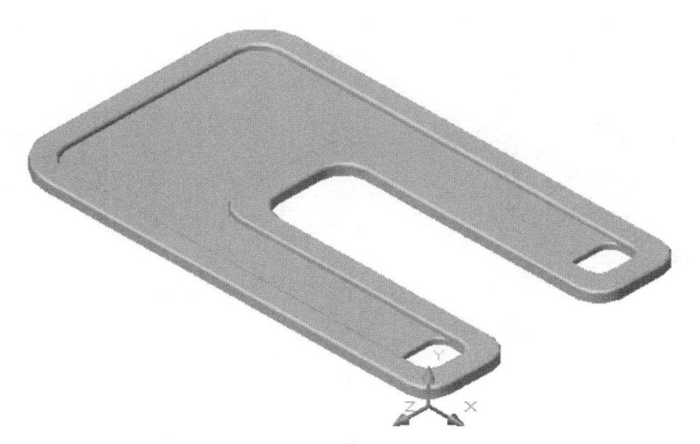

Figure 2–45

3. Place a sketch plane on the top face of the part.

4. Press the 9 key and press ENTER to look directly on the sketch view.
5. Sketch, Profile and Constrain a single line segment as shown in Figure 2–46.

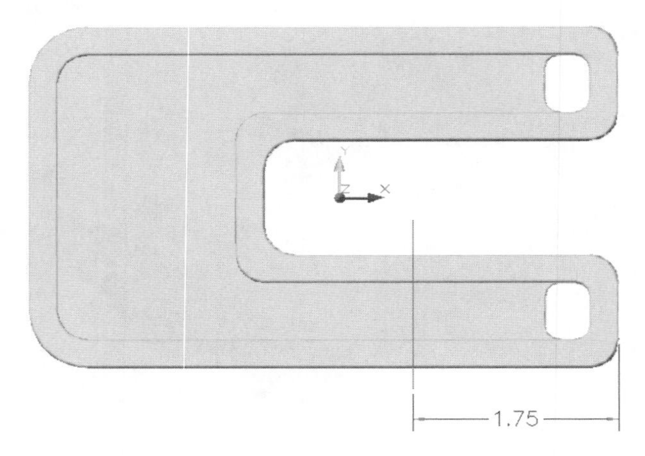

Figure 2–46

6. Press the 8 key to restore an isometric view
7. Right click on OpenProfile1 in the Browser and select BEND.
8. In the Bend dialog box, enter the values shown in Figure 2–47. Make sure the preview matches and click OK.

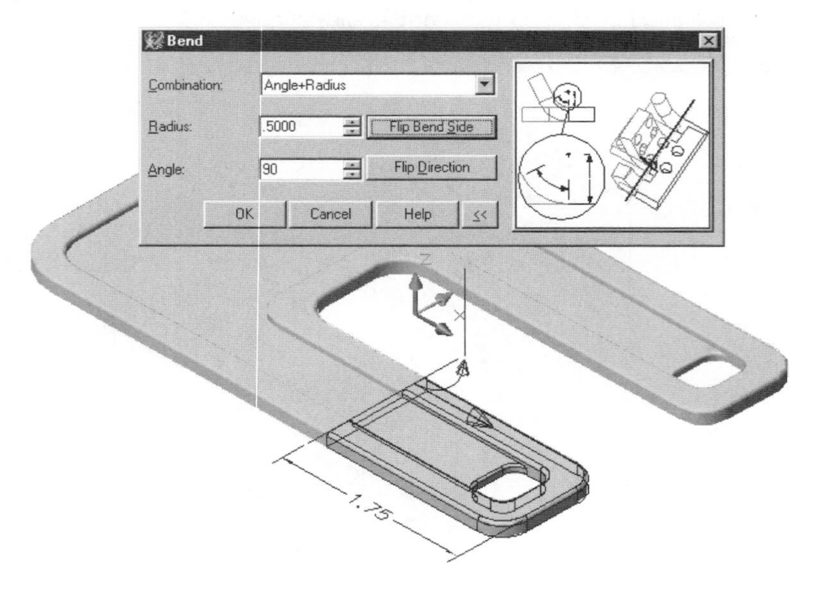

Figure 2–47

9. Right click on Bend1 in the Browser and select Edit Sketch. Add a vertical dimension of 3.5 to the line segment as shown in Figure 2–48.

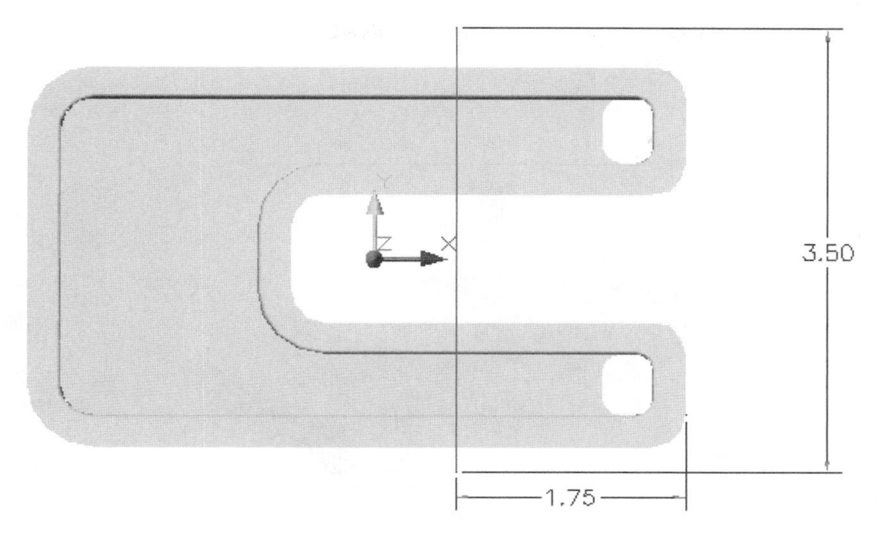

Figure 2–48

10. Right click in the graphics area and select Update Part.
11. The completed part should resemble Figure 2–49

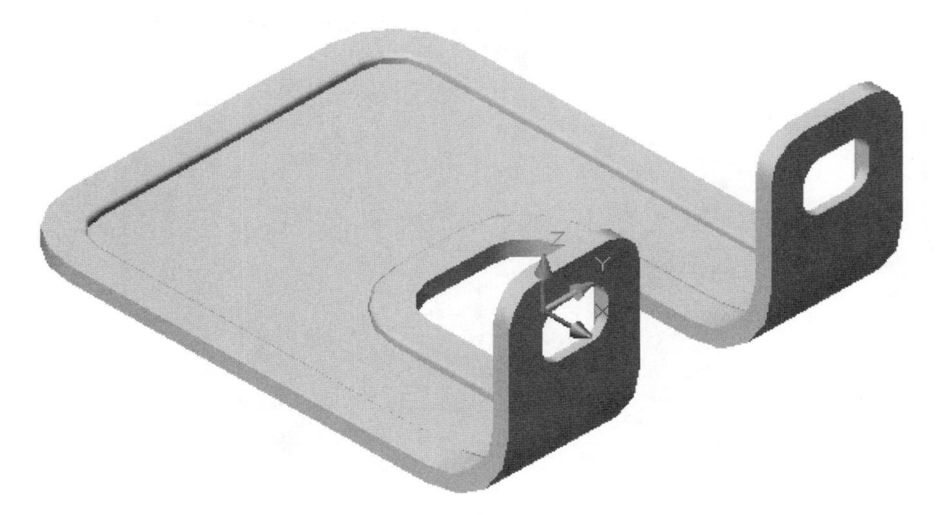

Figure 2–49

Tutorial 2–3 Open Profiles and Thin Extrusion Features

1. Start Mechanical Desktop 5 if it is not already running.
2. Open file MD02–EX 3.dwg. The file should resemble Figure 2–50. The current sketch plane is on the work plane.

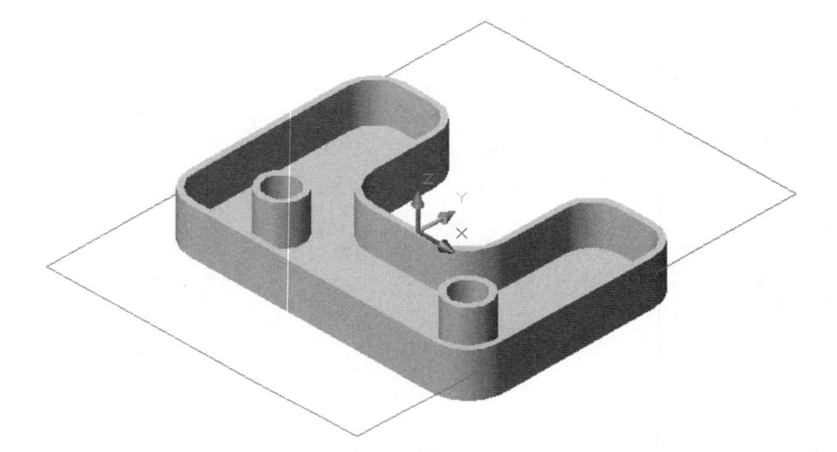

Figure 2–50

3. Press the 9 key and press ENTER to look directly on the sketch view
4. Sketch a shape that resembles Figure 2–51 and profile. Apply X Value constraint to the vertical lines to the center of the holes.

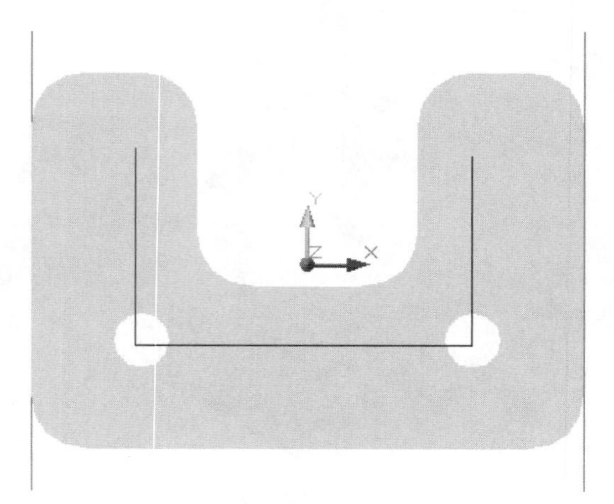

Figure 2–51

5. Access the AMEXTRUDE command and set the values as indicated in Figure 2–52. Make sure the arrow is pointing into the part. Click OK.

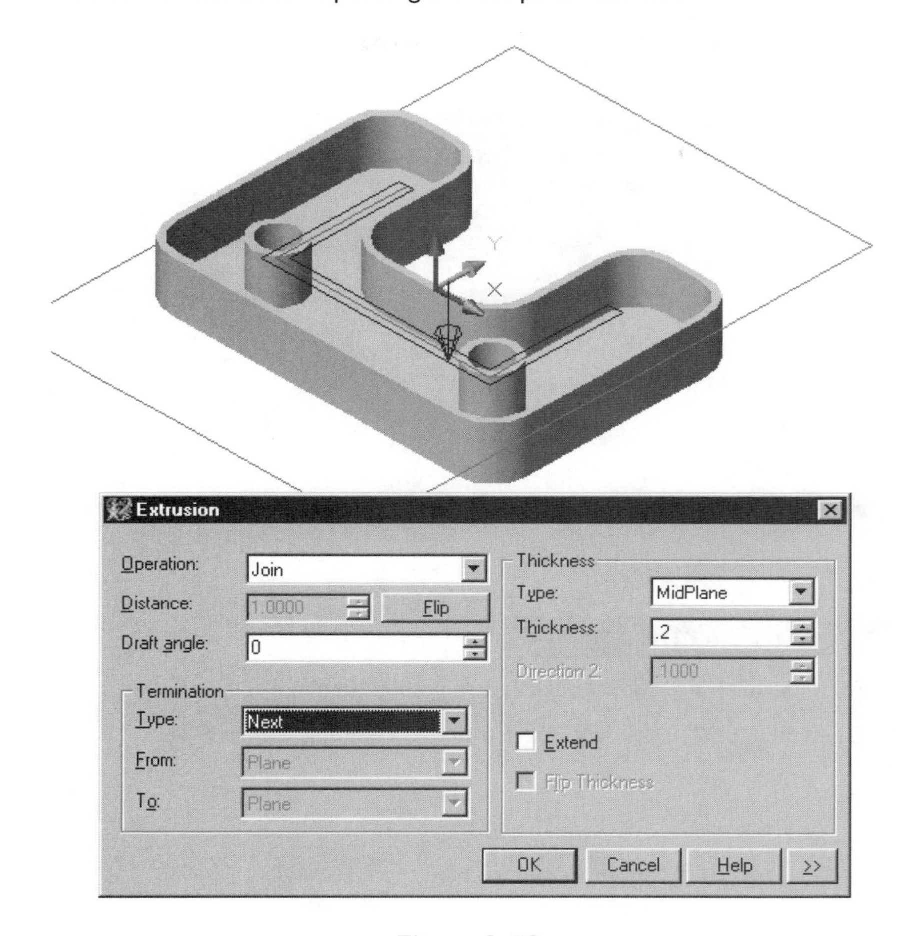

Figure 2–52

6. Use the 3D ORBIT command to rotate the view so that you can see that the thin extrusion did not go through the holes in the part. Also, notice that the extrusion does not reach the outer edge of the part.

7. Right click on ThinExtrusionToNext1 in the Browser and select Edit.

8. In the Extrusion dialog box, place a check mark in the box next to Extend and click OK and then press ENTER.

9. Right click in the graphics area and select Update Part. The thin extrusion now extends the sketch until it terminates into the part. See Figure 2–53.

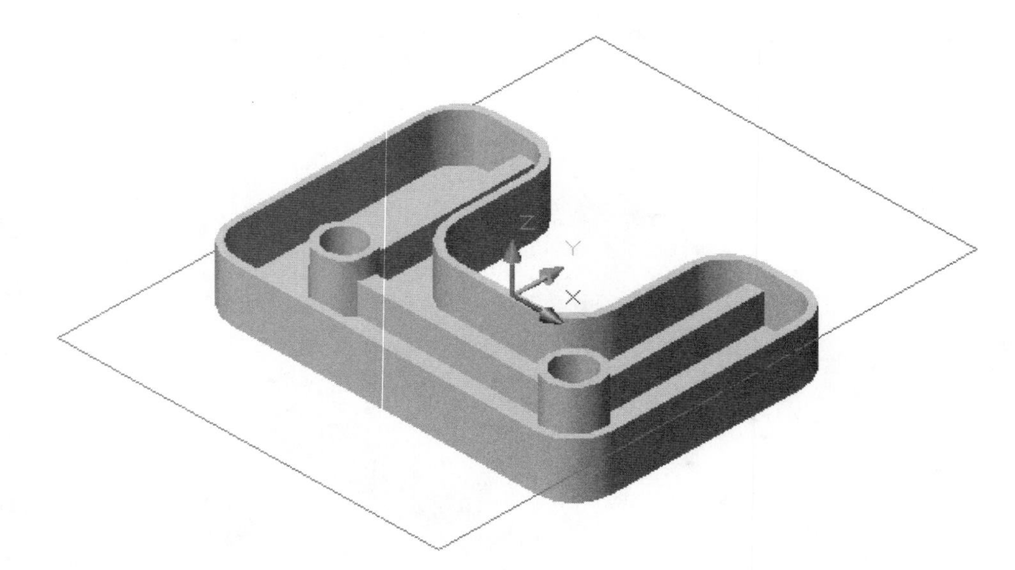

Figure 2–53

TUTORIAL

Tutorial 2–4 Pattern Features

1. Start Mechanical Desktop 5 if it is not already running.
2. Open file MD02–EX 4.dwg. The file should resemble Figure 2–54.

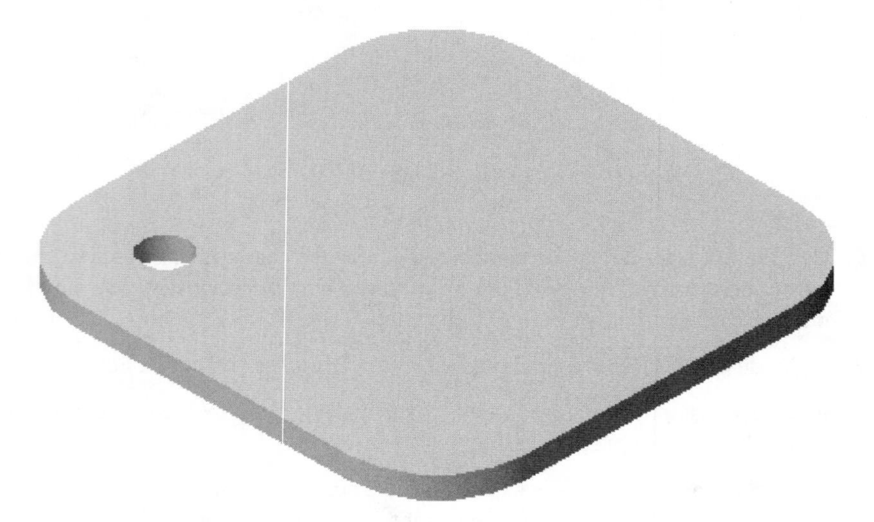

Figure 2–54

3. Right click on Hole1 in the Browser and select Pattern>Rectangular. In the Pattern dialog box, make the settings match those shown in Figure 2–55. Toggle between Incremental and Included Spacing and toggle the Flip Column/Row Direction button for both Column and Row Placement. Notice the changes in the preview as changes are made in the dialog box. If the preview does not change, click the arrow to the right of the Preview button and make sure both True and Dynamic are checked. Redo the settings to match Figure 2–55 and click OK. Your drawing should resemble Figure 2–56.

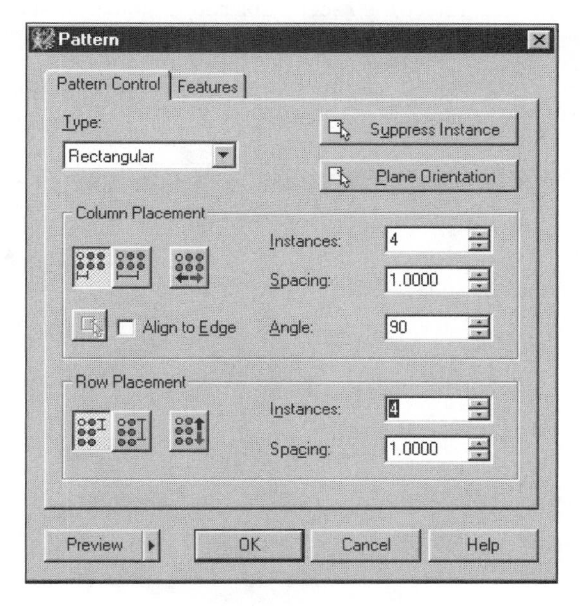

Figure 2–55

Figure 2–56

4. Place a .09 chamfer on the Hole1 feature.
5. Place the chamfer before Rectangular Pattern 1; you can do so by dragging Chamfer1 above Rectangular Pattern 1 in the Browser.
6. Right click on Rectangular Pattern 1 in the Browser and select Edit. The Pattern dialog box will display.
7. In the Pattern dialog box select the Features tab and then click the Add button.
8. In the graphics area, select the recently placed chamfer. Be sure that the chamfer is selected and not ExtrusionBlind1.
9. Press ENTER (or right click) to return to the Pattern dialog box. You should see both Hole1 and Chamfer1 listed. See Figure 2–57.

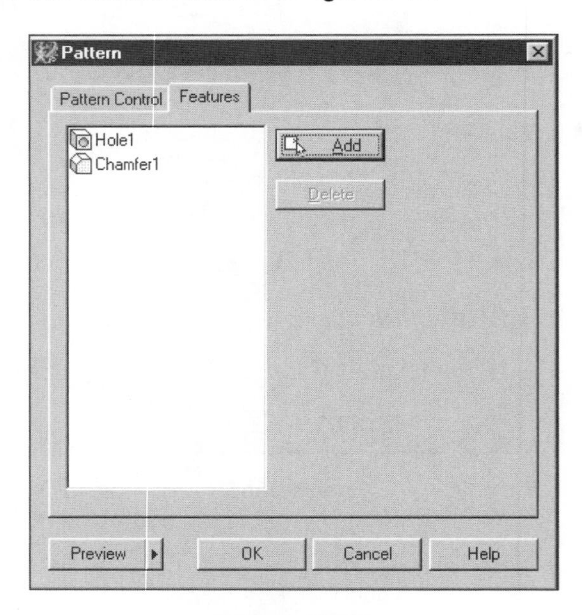

Figure 2–57

10. Click OK and the chamfer should now be on all the holes.
11. Right click on Rectangular Pattern 1 in the Browser and select Edit. The Pattern dialog box should display.
12. Place a check mark in the box next to Align to Edge. Select the button just to the left of the check mark and when prompted, click the vertical edges of the part and then click OK.
13. Update the part. At this point, the appearance of the part should not change.
14. Press the 9 key and press ENTER to look directly on the sketch view.

TUTORIAL

15. Access Design Variables. There should be one variable called Angle in the Design Variables dialog box. You can access Design Variable from Part > Design Variables.

16. Double-click in the equation field when you've located the 90 value. Change 90 to 110 and click OK.

17. The part should automatically update and resemble Figure 2–58.

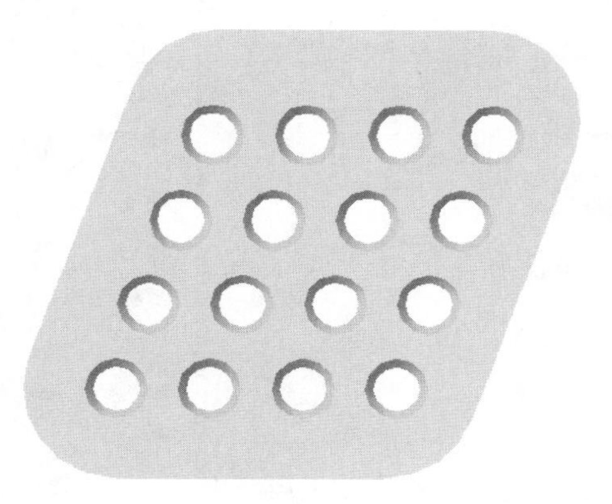

Figure 2–58

TUTORIAL

Tutorial 2–5 Axial Features

1. Start Mechanical Desktop 5 if it is not already running.
2. Open file MD02–EX 5.dwg. The file should resemble Figure 2–59.

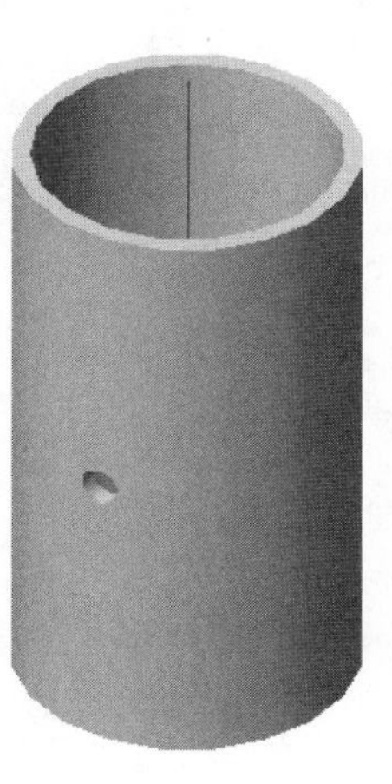

Figure 2–59

3. In the Browser, right click on ExtrusionThru1 and select Pattern > Axial.
4. At the prompt for Select Rotational Center, select the Work Axis on screen.
5. In the Pattern dialog box, match the settings and preview to those shown in Figure 2–60. Click OK.

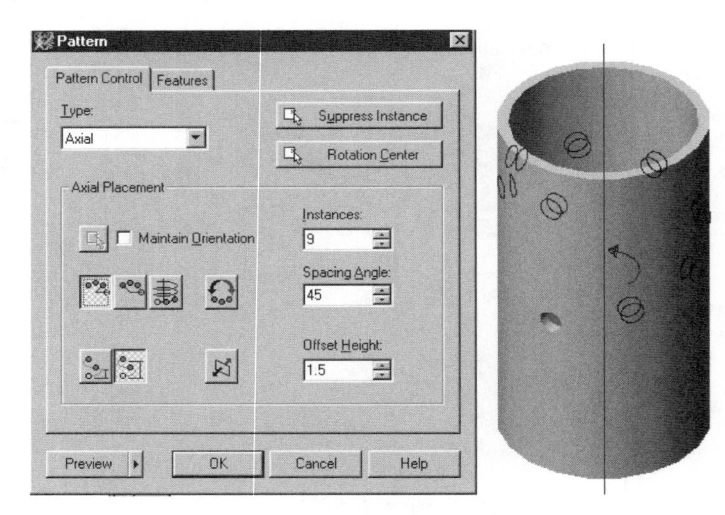

Figure 2–60

6. Repeat Steps 3, 4 and 5. Use the exact same setting in the Pattern dialog box, except this time select the Flip Offset Direction button, shown in Figure 2–61.

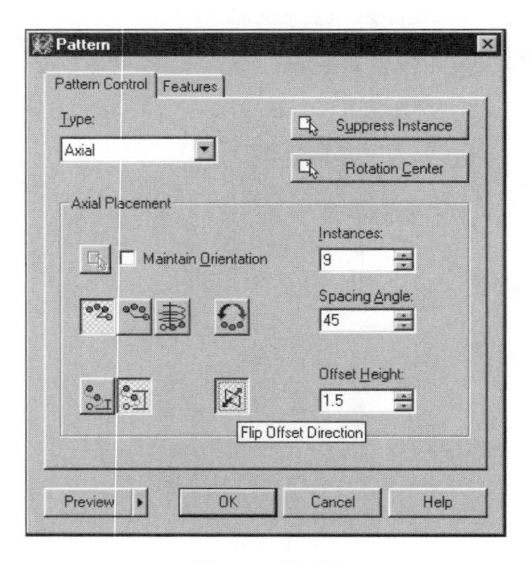

Figure 2–61

7. Click OK. The pattern will complete and your file should resemble Figure 2–62.

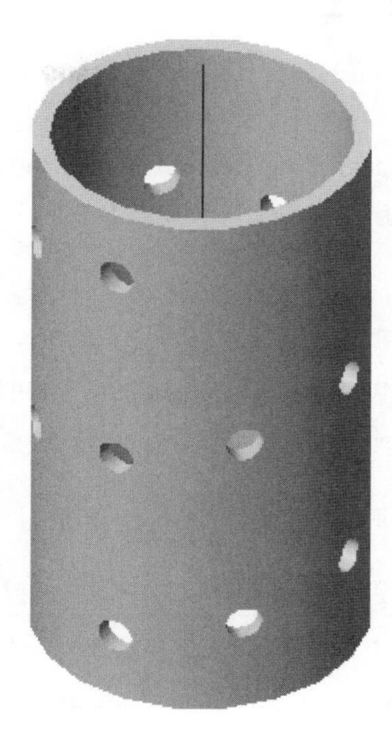

Figure 2–62

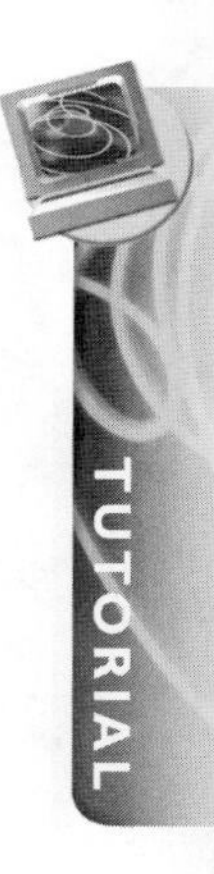

REVIEW QUESTIONS

1. Any True Type font can be used for Text Sketch and Text Features. T or F?
2. Any combination of lines, arcs, polylines and splines can be used for Bend open profiles. T or F?
3. Bend sketches must lie on the part. T or F?
4. You must select the type of pattern to be created before entering the Pattern dialog box. T or F?
5. Thin extrusions fill in any gaps encountered. T or F?
6. Rib features are created ______________________________ to the sketch plane.
7. Thin extrusions are created ______________________________ to the sketch plane.
8. Rib features will fail unless all sections of the sketch extend to intersect the part. T or F?
9. Features can now be added to Pattern features after the Pattern feature has already been created. T or F?
10. Rectangular Pattern features can use existing part edges to parametrically control the pattern angle. T or F?

Assembly Modeling Enhancements

Mechanical Desktop 5 incorporates many new tools to increase productivity. The ability to edit in place has been enhanced with the ability to edit external sub-assemblies and combine parts without localizing. With this new capability, you can move one or multiple components from the main assembly into any local or external subassembly and maintain assembly constraints on those components. The ability to select multiple features, described in Chapter 1, also allows the use of drag and drop functionality to select and restructure multiple parts.

After completing this chapter, you will be able to:

- Move Single or Multiple Parts to Local or External Subassemblies
- Understand and Use the AMRESTRUCTURE Command
- Drag And Drop Parts Into Assemblies
- Understand How Constraints are Handled When Restructuring Assemblies
- Reorder Parts in Assemblies
- Understand Local and External Subassemblies Visibility

MOVING PARTS BETWEEN LOCAL AND EXTERNAL SUBASSEMBLIES

Parts can be added to assemblies in a number of different ways. They can be created locally, attached through the catalog as a single part or as part of a subassembly. At times, it might be necessary to move parts from one subassembly to another. Mechanical Desktop 5 allows you to do this without losing assembly constraints. Mechanical Desktop 5 takes a part from either the host assembly file or an external subassembly and moves that part into another subassembly maintaining current assembly constraints.

To perform these tasks, Mechanical Desktop 5 uses a new command called AMRESTRUCTURE. This new command works only on the command line and therefore could quickly become a tedious command to use. Fortunately, you can accomplish the restructuring of assemblies much more efficiently using the Browser. You select the part or parts that you want to restructure and then drag and drop them into the new assembly. The parts selected can originate either in the host assembly file or in local or external subassemblies. The destination of the restructured parts can also be a host assembly file or in local or external subassemblies. To select multiple parts, hold down the CTRL key and pick on the parts to be moved in the browser. Once selected, simply drag and drop the parts into the new location. Mechanical Desktop 5 prompts where it would be appropriate to drop the parts. Figure 3–1 is an example of the Browser displaying the current structure of the assembly.

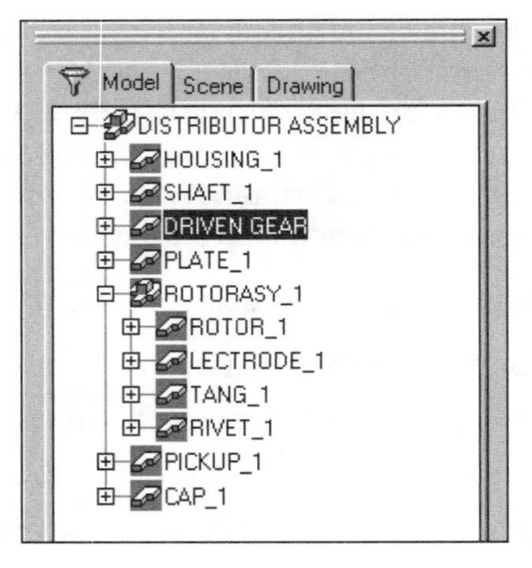

Figure 3–1

Suppose you want to create a new subassembly for the distributor shaft and also move the driven gear into that subassembly. Figure 3–1 shows that the shaft and gear are external parts. To create the new subassembly, right-click in the Browser and select New Subassembly. At the command line, enter the name of the new subassembly and press ENTER. Then, holding down the CTRL key select both the shaft and the driven gear and drag them down to the new subassembly. Mechanical Desktop 5 displays the universal "NO" symbol (a circle with a diagonal line through it) if the cursor is located where it would be inappropriate to place the selected parts, such as on another part. Once a valid location is found, the cursor changes to indicate that the items can be dropped. Figure 3–2 shows the selected parts being placed in the new subassembly.

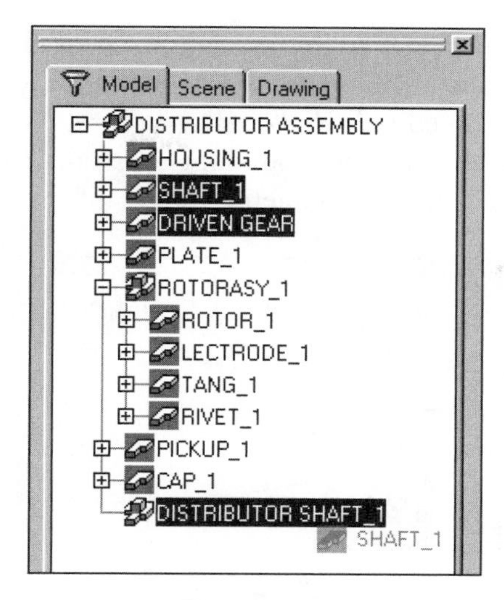

Figure 3–2

HOW CONSTRAINTS ARE HANDLED WHEN RESTRUCTURING ASSEMBLIES

In the Browser, notice that Driven Gear and Shaft_1 have been moved into subassembly Distributor Shaft_1. In addition to the parts, all the constraints on them have moved as well. In Figure 3–3 look at the first constraint. Notice the small yellow triangle to the left of the Mate plane-to-plane constraint icon.

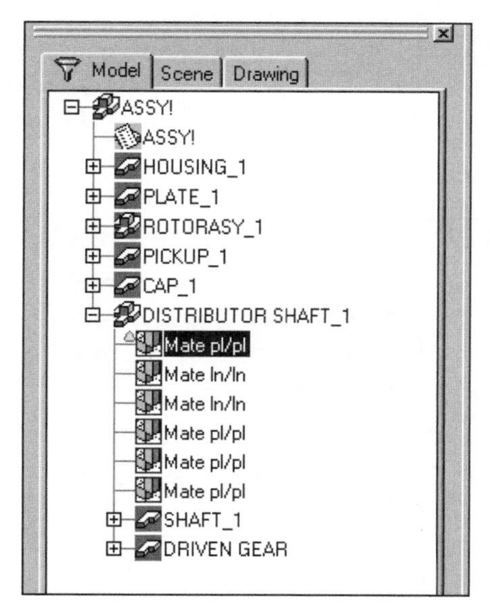

Figure 3–3

The yellow triangle indicates that the restructuring of these parts into the new subassembly created a redundant constraint. Redundant constraints are not used in the assembly and cannot be edited. Redundant constraints can only be deleted. As seen in Figure 3–4, right-clicking the redundant constraint in the Browser shows the only available option on the menu is Delete.

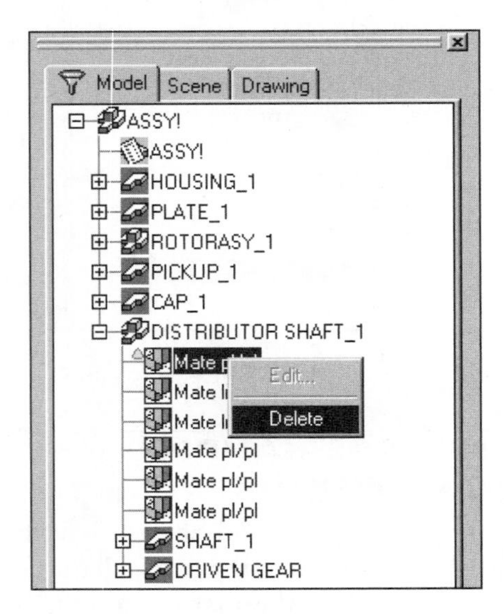

Figure 3–4

To use AMRESTRUCTURE at the command line requires typing in the exact names of the parts and subassemblies involved. This could quickly become a very tedious operation. Once complete, externalize the parts and Mechanical Desktop will update both of the external files.

HOW TO REORDER PARTS IN ASSEMBLIES

You have seen how to use the CTRL key to select multiple parts and restructure assemblies. When using the CTRL key, Mechanical Desktop assumes that you want to restructure the assembly. However, you might still need to reorder the parts and subassemblies within the current assembly file.

To reorder parts within the current assembly, use the SHIFT key instead of the CTRL key. Hold down the SHIFT key and select the desired part to reorder and then drag it into its new location.

LOCAL AND EXTERNAL SUBASSEMBLIES VISIBILITY

So far, many ways to edit both local and external subassemblies have been demonstrated. In addition to these new tools, Mechanical Desktop also provides new visualization tools. Mechanical Desktop offers several different options to view parts and assemblies. For example, Figure 3–5 shows an assembly in the normal gouraud shaded mode.

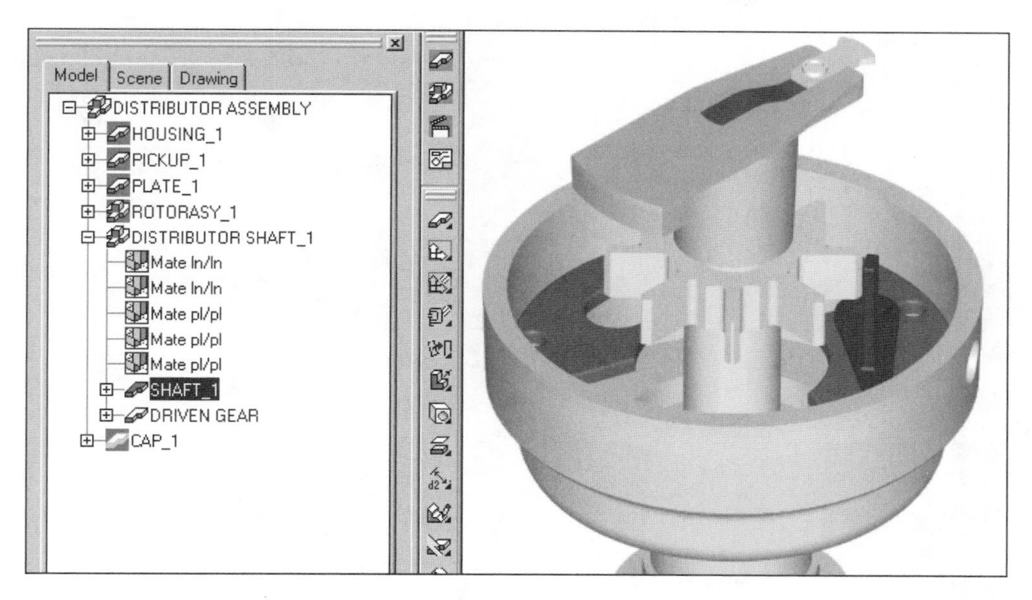

Figure 3–5

Figure 3–6 shows that when a subassembly is opened for editing, and the current visualization selection is 3D wireframe, or any of the shaded options, then the remainder of the assembly is grayed out. This makes it possible to see the rest of the assembly, but still work easily with the items to be edited. This display is constant for both local and external assemblies. While Figure 3–6 illustrates a local assembly open for editing, Figure 3–7 displays an external assembly open for editing.

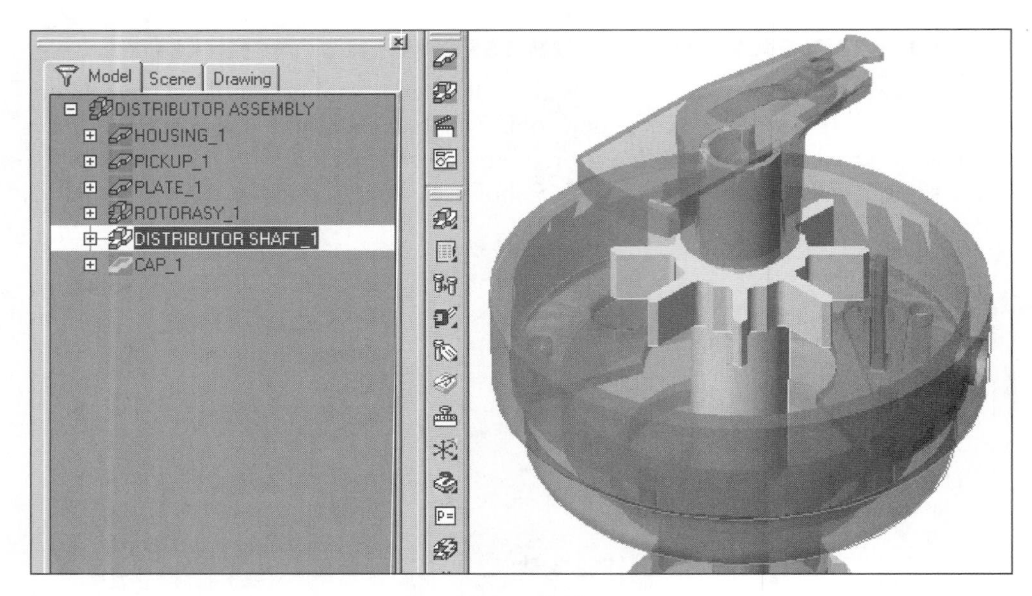

Figure 3–6

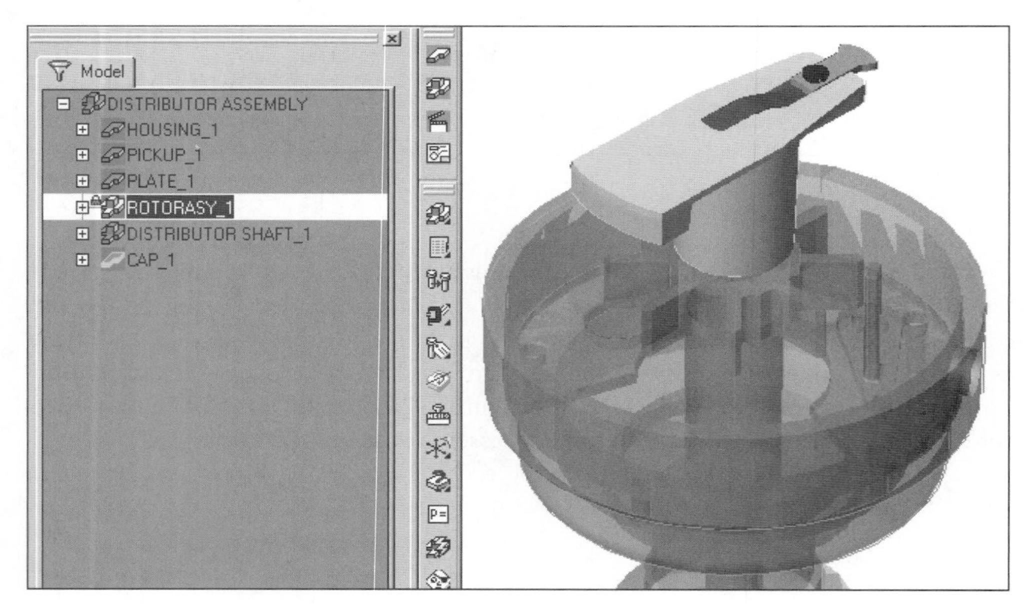

Figure 3–7

Tutorial 3–1 Assembly Modeling Enhancements

1. Start Mechanical Desktop 5 if it is not already running.
2. Open file MD03-EX1.dwg.

   The file should resemble Figure 3–8. This is a very basic reservoir assembly; many of the features and parts have been omitted for clarity. For this assembly, all parts are local and have been created with no subassemblies.

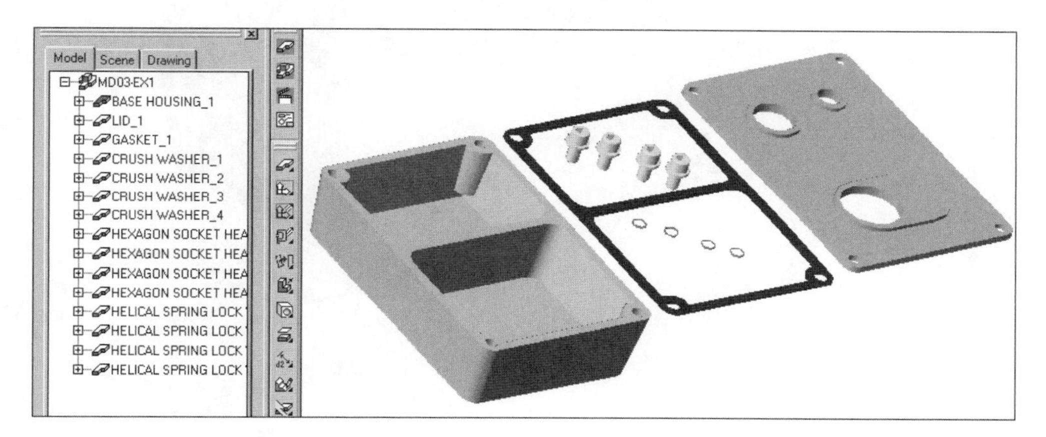

Figure 3–8

The first thing to do is organize the assembly in a logical manner. You will use AMRESTRUCTURE to rearrange the parts in this assembly to a logical order with both local and external subassemblies.

3. Press the 8 key and then ENTER to go to a true isometric view.
4. Assemble the file by updating the assembly. Use the hot key option of "aa" and press ENTER or use the Update Assembly icon at the bottom of the Browser. Once assembled, the file should resemble Figure 3–9.
5. Right-click an open area of the Browser and select New Subassembly. Type FASTENER SET on the command line. Fastener Set_1 displays in the Browser.
6. Hold the CTRL key down and select all the Hexagon Sockets and Helical Springs listed in the Browser.

Figure 3–9

7. Once all are selected, drag and drop the selected items into Fastener Set_1.
8. Double click Fastener Set_1 in the Browser to open it for editing. The file should resemble Figure 3–10. Notice how the parts not in Fastener Set_1 are grayed out. Also, notice the redundant constraints in the Browser.

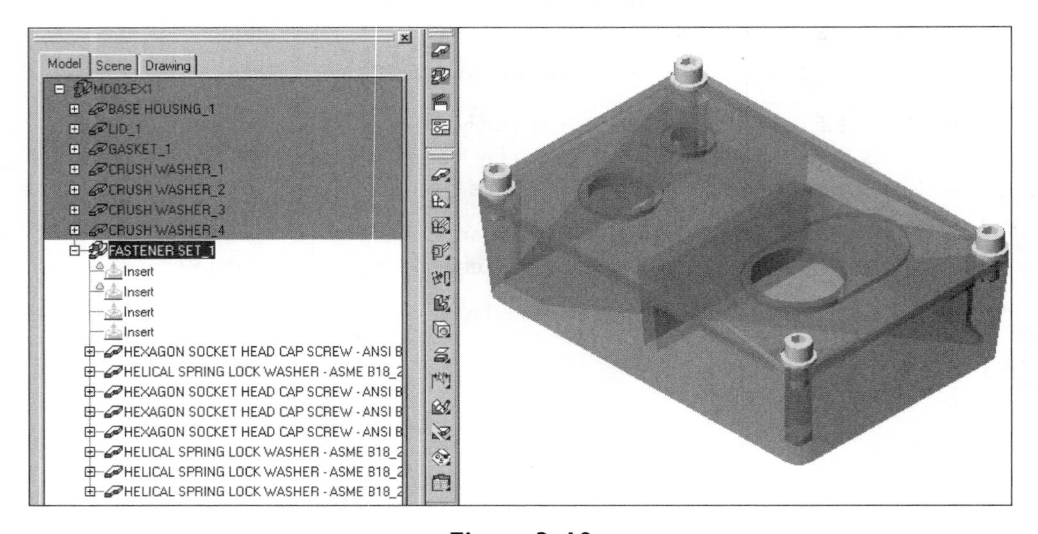

Figure 3–10

9. Activate the Main Assembly by double-clicking MD03-EX1 at the top of the Browser.
10. Delete the redundant constraints by right-clicking them and selecting Delete.
11. Repeat Step 5 and create a subassembly called Gasket Assembly.
12. Select Gasket_1 and all Crush washers in Gasket Assembly_1.
13. Access the Mechanical Desktop Catalog. Under the Local Assembly Definitions, right-click Gasket Assembly and select Externalize.
14. Double-click Gasket Assembly in the Browser and observe the graphics area.
15. Activate the Main Assembly by double-clicking MD03-EX1 at the top of the Browser.
16. Hold down the SHIFT key, select LID_1 and drag it above the base house.

REVIEW QUESTIONS

1. The AMRESTRUCTURE dialog box makes it easy to restructure assemblies. T or F?
2. Parts can only be restructured with local subassemblies. T or F?
3. Once restructured, all parts lose their previous assembly constraints and must be restructured. T or F?
4. Redundant constraints can be edited. T or F?
5. A yellow triangle in the Browser indicates the constraints that need to be edited. T or F?
6. The CTRL key is now used to reorder parts. T or F?
7. When a subassembly is opened for editing, all parts automatically gray out, regardless of the shading. T or F?
8. Only local assemblies gray out when edited. T or F?
9. Using AMRESTRUCTURE at the command line is the most efficient way to restructure assemblies. T or F?
10. The SHIFT key is now used to reorder parts in an assembly. T or F?

CHAPTER 4

Drawing Manager Enhancements

Several new enhancements have been added to drawing views in Mechanical Desktop 5. Enhancements have been made in the Creation of Views, Annotations, Dimensions, and exporting of views, just to name a few. Mechanical Desktop 5 has added the ability to place reference dimensions to arc-shaped splines, and Power dimensions to support Endpoint, Midpoint, Center, Quadrant, and Apparent Intersection object snaps. Improvements have been made to Section View and Detail View labels.

After completing this chapter, you will be able to:

- Select the Active Part in Scene Mode
- Dimension To Arc-Shaped Splines
- Power Dimension with Object Snaps
- Move Multiple Dimensions
- Create Non Rectangular Viewports for Detail Views
- Export Views To Multiple Formats

ACTIVE PART SELECTION IN SCENE MODE

Mechanical Desktop 5 now allows selection of the active part while creating drawing views in files that contain scenes. Previously, the select option was the only way to access the active part. Now, in the Create Drawing View dialog box, you can specify Active Part for the drawing view, as shown in Figure 4–1.

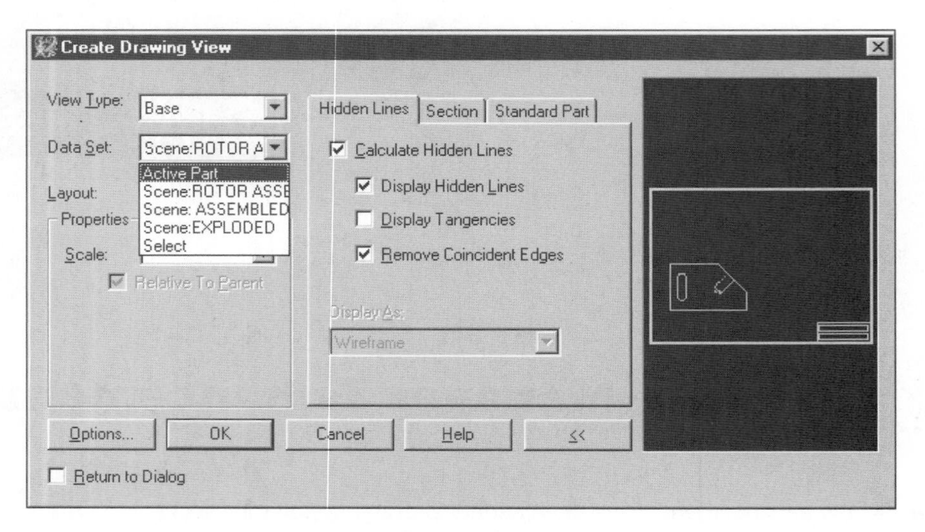

Figure 4–1

DIMENSION TO ARC-SHAPED SPLINES

Mechanical Desktop 5 now allows reference dimension to arc-shaped splines and tapped holes. Arc-shaped splines are most commonly created when a round shape is cut through another round shape. An example of this is drilling a hole through a piece of bar stock perpendicular to the centerline of the bar stock. The location where the drill hole and the outer edge of the bar stock would be, is represented by an arc-shaped spline in the drawing views, as illustrated in Figure 4–2.

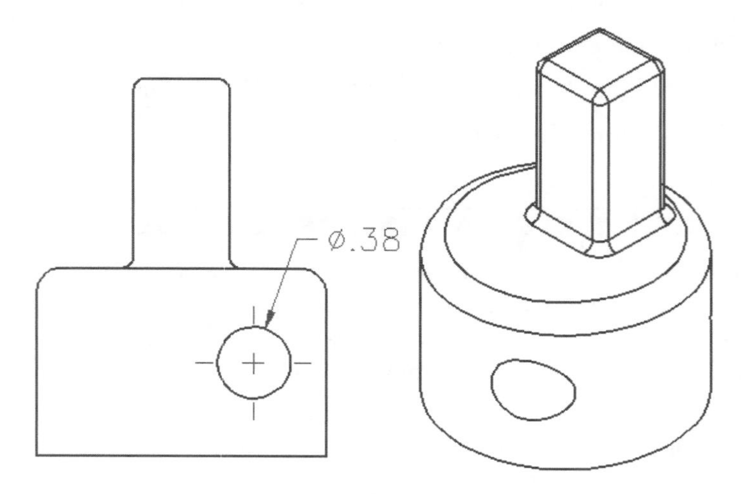

Figure 4–2

POWER DIMENSION WITH OBJECT SNAPS

In previous releases of Mechanical Desktop, it was often frustrating to dimension to the outer edges of circular shapes. Mechanical Desktop would always default to the center of the arc or circle. This was extremely bothersome when trying to dimension the ends of shafts. Mechanical Desktop 5 resolves this problem by allowing Osnaps (Object Snaps) to be used when placing reference dimensions (see Figure 4–3). Osnaps are supported only when using Power Dimensioning to add reference dimensions in drawing views.

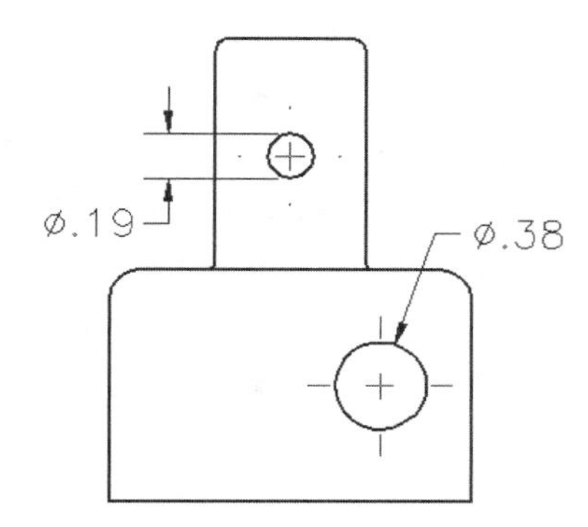

Figure 4–3

MOVE MULTIPLE DIMENSIONS

The AMMOVEDIM command has a new option added to it. Now you have the option to move multiple dimensions in a single use of the command. When the AMMOVEDIM command is accessed, the new option, move mUltiple, appears in the command line (see Figure 4–4).

Figure 4–4

You can select the move mUltiple option by entering the letter U at the command line. Or you can right click in the graphics area and select "move mUltiple" from the context menu, as shown in Figure 4–5. Once the option has been selected, Mechanical Desktop prompts you to "Select dimensions to move" and remains at that prompt until you press the ENTER key, or right click to indicate you are done selecting. In Figure 4–6, the .31 and .63 dimensions will be selected so they can be moved to the view below.

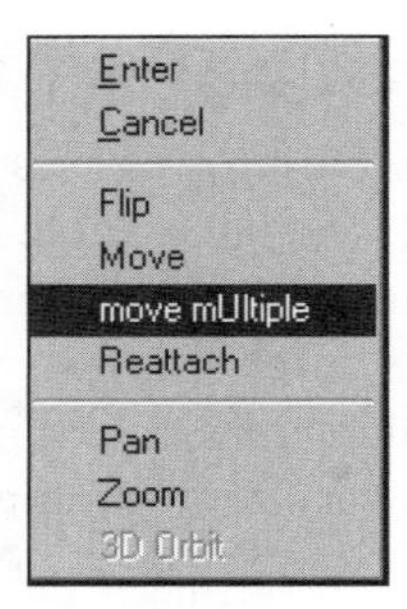

Figure 4–5

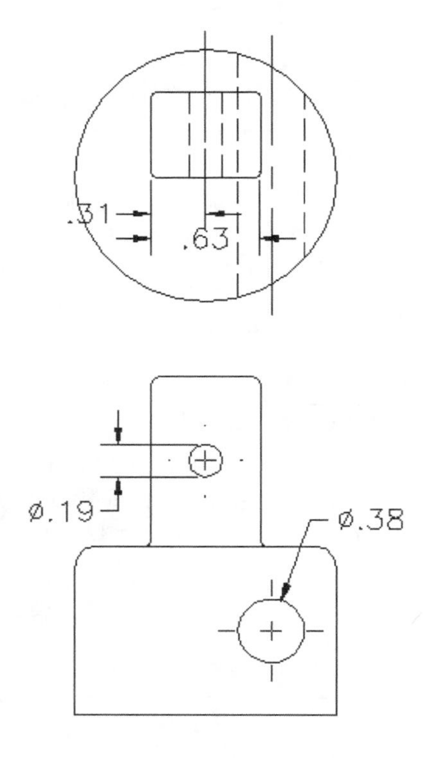

Figure 4–6

After selecting the dimensions, Mechanical Desktop prompts you to "Select the destination view:" Select the view that you want to move the dimensions to. Mechanical Desktop will move the dimensions into that view, provided they are appropriate to the new view (see Figure 4–7). If Mechanical Desktop encounters dimensions that are not appropriate, a message will be displayed at the command line stating: "Some dimensions could not be moved to designated view" and the dimension will not appear.

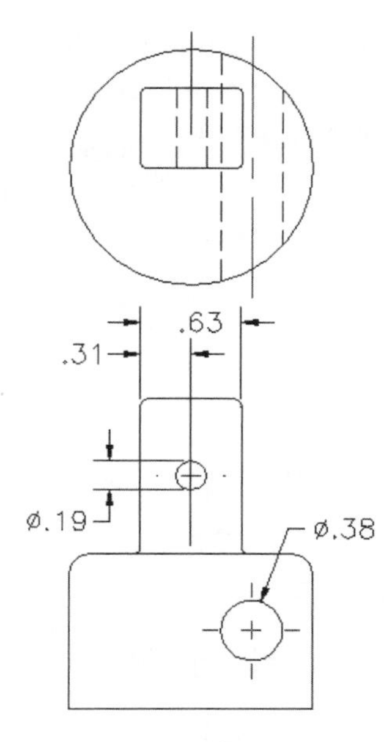

Figure 4–7

VIEWPORT ENHANCEMENTS

In Mechanical Desktop 5, you can now create non-rectangular viewports and resize viewports. In previous releases viewports could be resized, but Mechanical Desktop always returned them to their original size when the drawing views were updated. And, in the case of Section or Detail view, if the labels were edited, they also reverted back to their original format. In Mechanical Desktop 5, viewports and label edits remain in their edited state after view updates.

CREATE NON-RECTANGULAR VIEWPORTS FOR DETAIL VIEWS

New options for the creation of viewports appear when creating a detail view. The process for creating a detail view remains the same. When Mechanical Desktop prompts you to specify the center point for the detail view, new options appear at the command line for creating the view (see Figure 4–8). The new options can be accessed by entering the capital letter of the option, or right clicking in the graphics area and selecting the option from the context menu that appears.

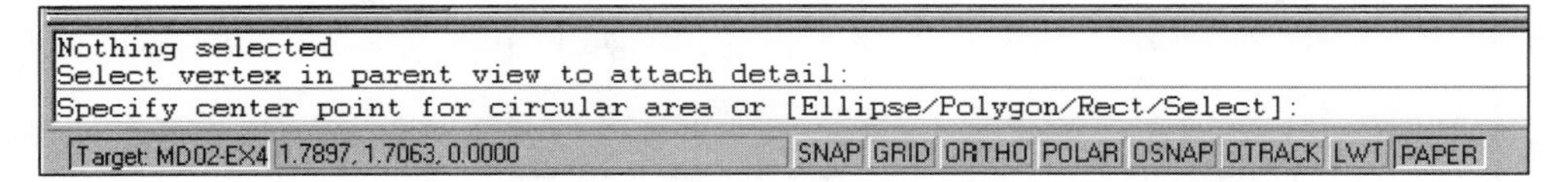

Figure 4–8

There are four different options for creating the detail view: Ellipse, Polygon, Rect (Rectangle) and Select. Ellipse and Rectangle are preset shapes that require input for size only. Polygon and Select are different.

When using the Polygon option, you define the shape that is going to be displayed. Figure 4–9 shows the detail view being created and Figure 4–10 shows the end result.

Note: The AMVIEWS layer was turned on to show the shape created.

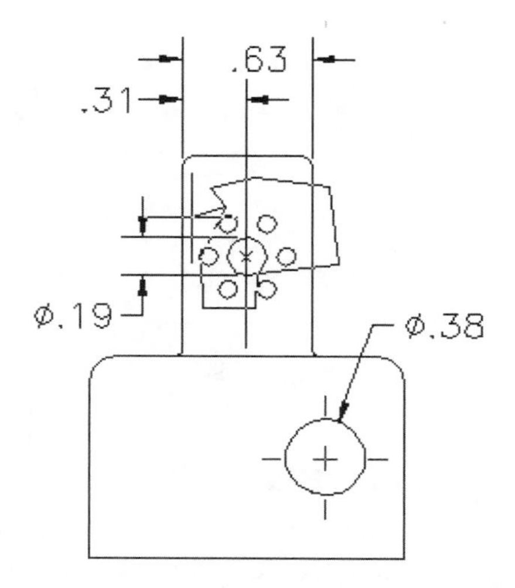

Figure 4–9

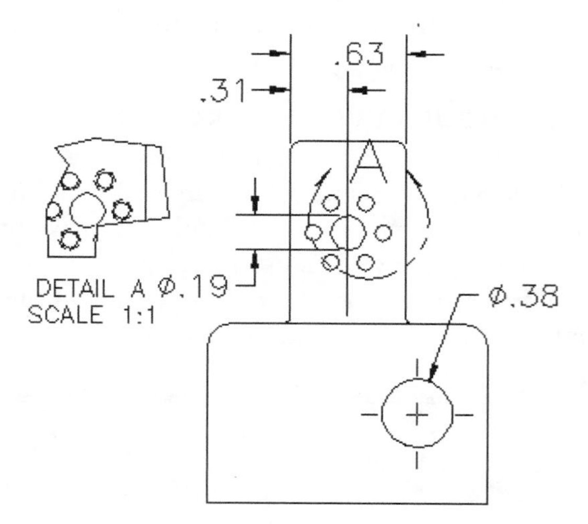

Figure 4–10

The Select option requires a predefined shape be created before the drawing view is made. In Figure 4–11, a closed spline entity was created around the same objects as the polygon view. When the Select option is used, Mechanical Desktop prompts: "Select closed clipping curve to define detail area." This means the shape selected must form a closed shape. Individual lines and arcs do *not* fulfill this requirement. However, the PEDIT or BPOLY commands can be run to convert the individual lines and arcs into a closed shape. Figure 4–12 shows the completed detail view using the Select option.

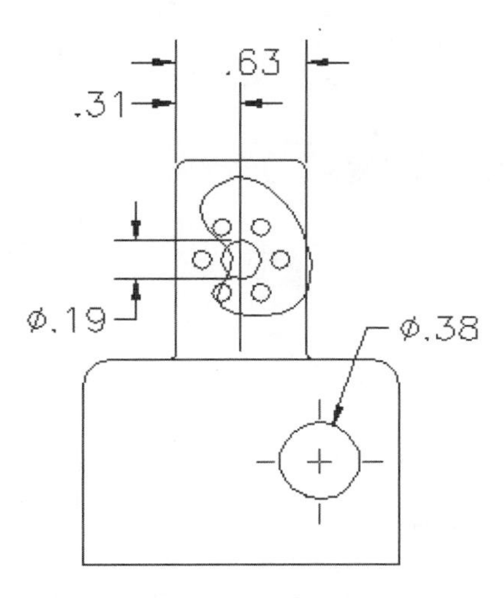

Figure 4–11

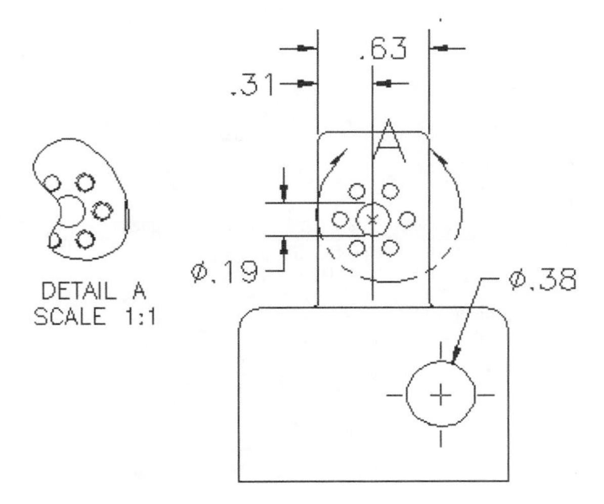

Figure 4–12

EXPORT VIEWS TO MULTIPLE FORMATS

Mechanical Desktop 5 now allows us to export views to multiple formats. Previously, you could export only the views to a .dwg format, based on the current release of AutoCAD. Now when you access the AMVIEWOUT command, you can select the format type to be created. Figure 4–13 shows the Export drawing views to AutoCAD 2d dialog box. At the bottom of the dialog box, select the desired format for output.

Individual views can be selected for output by right clicking on the specific view in the Mechanical Desktop Browser. Entire layouts can be exported by right clicking on the layout name and selecting Export all views. Regardless of the method used, the Export drawing views to AutoCAD 2d dialog box will appear.

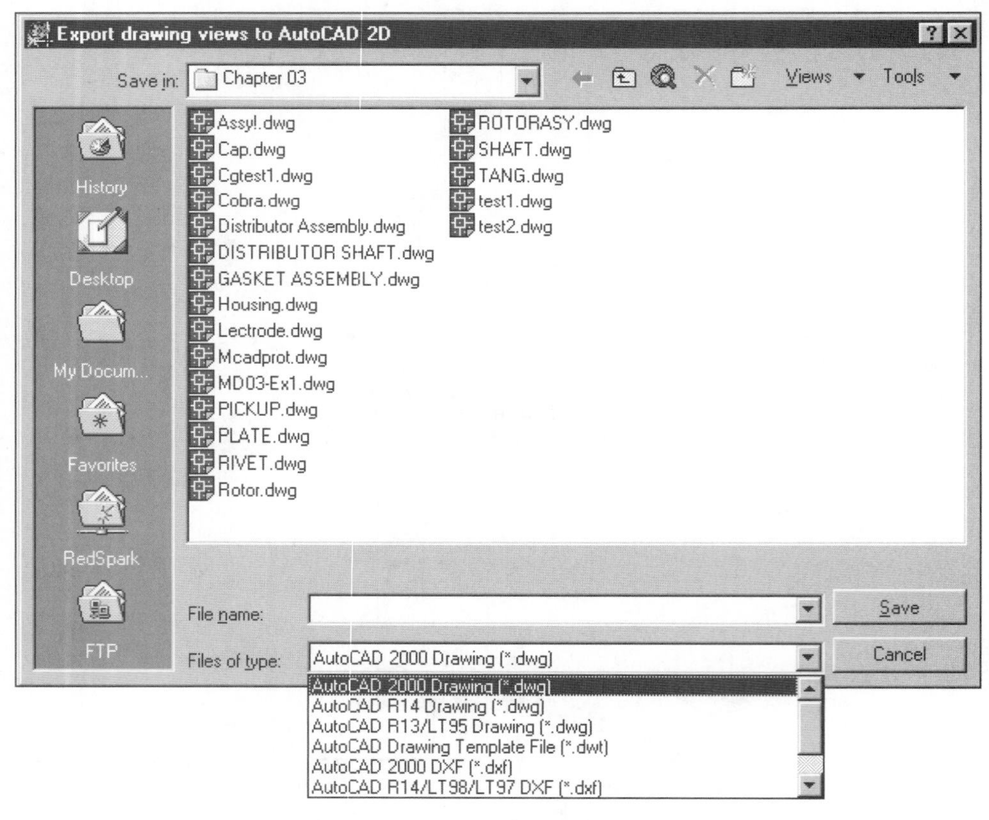

Figure 4–13

Tutorial 4–1 Drawing Manager Enhancements

1. If not already running, start Mechanical Desktop 5.
2. Open file MD04-EX 1.dwg. The file should resemble Figure 4–14.

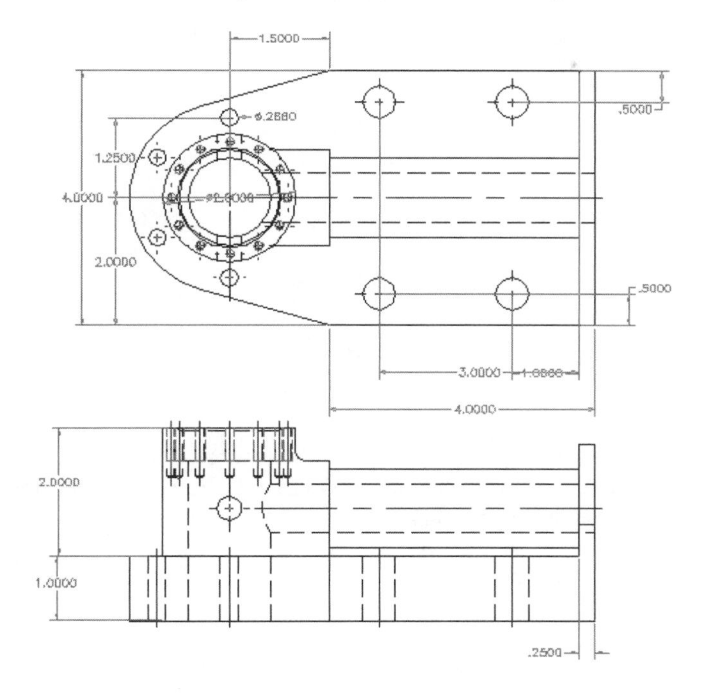

Figure 4–14

3. Zoom in on the bottom view until your screen resembles Figure 4–15

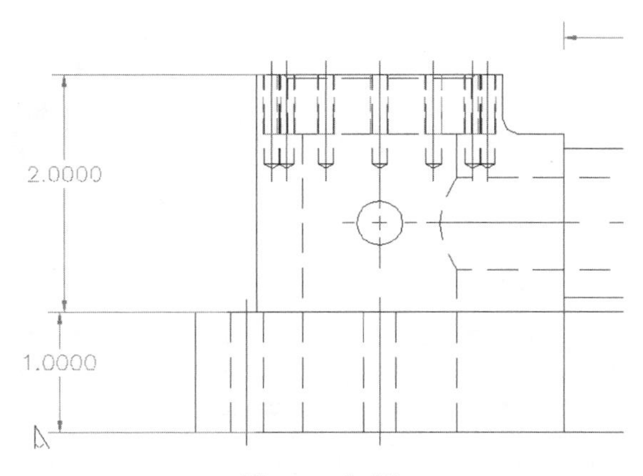

Figure 4–15

4. Access the AMREFDIM command and select the circle in the middle of the screen. That circle is actually a spline arc representing the circle. Place the reference dimension to the left, as shown in Figure 4–16.

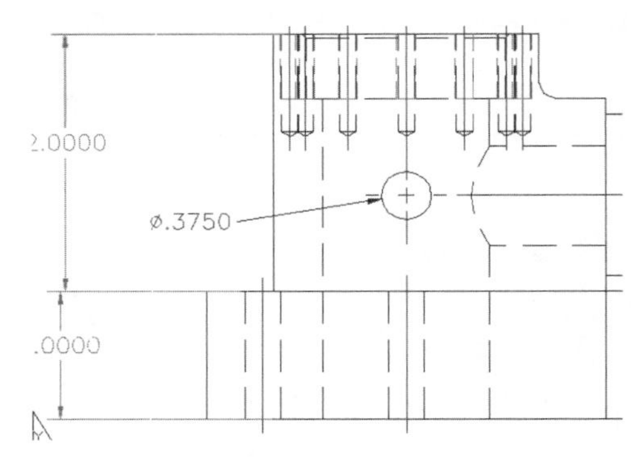

Figure 4–16

5. Pan up to the top view.

6. Access the AMREFDIM command and place a dimension using the quadrant Osnaps on the large circle. (This circle currently has a 2" diameter dimension.) Your result will resemble Figure 4–17.

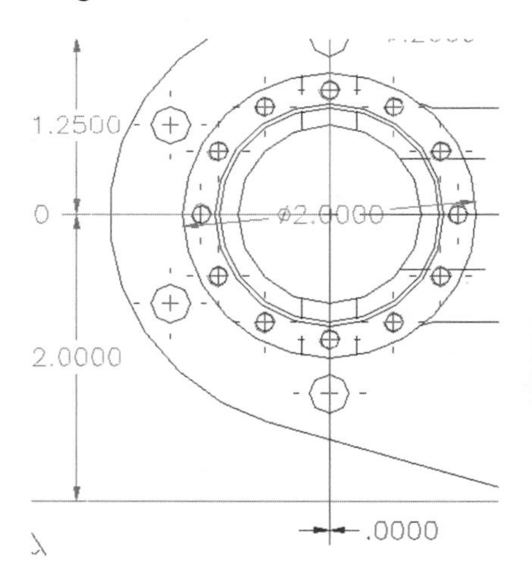

Figure 4–17

7. This time access the AMPOWERDIM command and place the same dimension, again using Osnaps. It may be helpful to pick on the circle just below the centerline to insure selection of the quadrants of the circle.

8. Place the dimension. In the Power Dimensioning dialog box, you can place the diameter symbol by adding %% to the front of the dimension as shown in Figure 4–18.

TUTORIAL

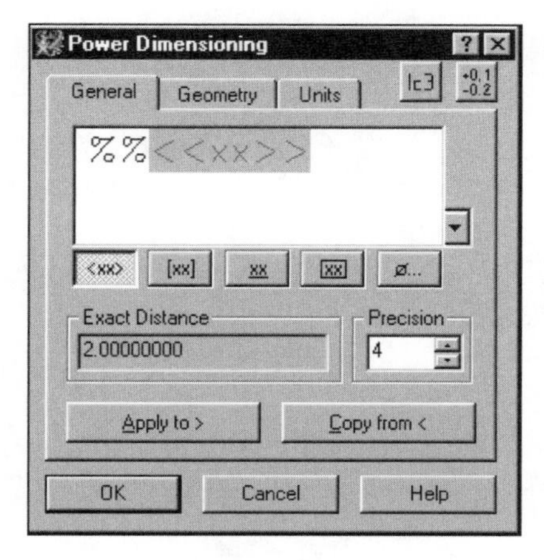

Figure 4–18

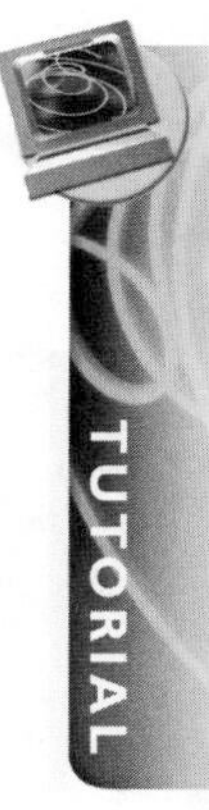

9. Perform a Zoom Extents.
10. Access the Move Dimension (AMMOVEDIM) command and select the move mUltiple option by right clicking in the graphics area.
11. Select the 2" diameter dimension just created and the 1.5000 dimension at the top of the top view.
12. Press the ENTER key.
13. When prompted for the destination view, select the bottom view.
14. Grip edit the dimensions and locate them in a suitable place. Depending on the placement of your dimensions, your drawing should resemble Figure 4–19.

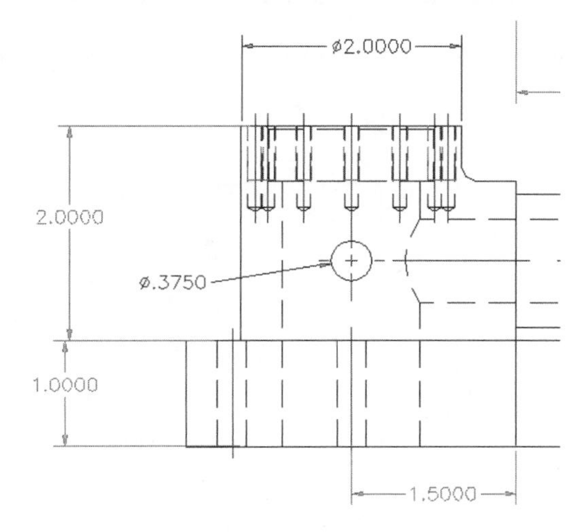

Figure 4–19

15. Zoom out and place an Ortho view to the right side of the existing Ortho view. Your drawing should resemble Figure 4–20.

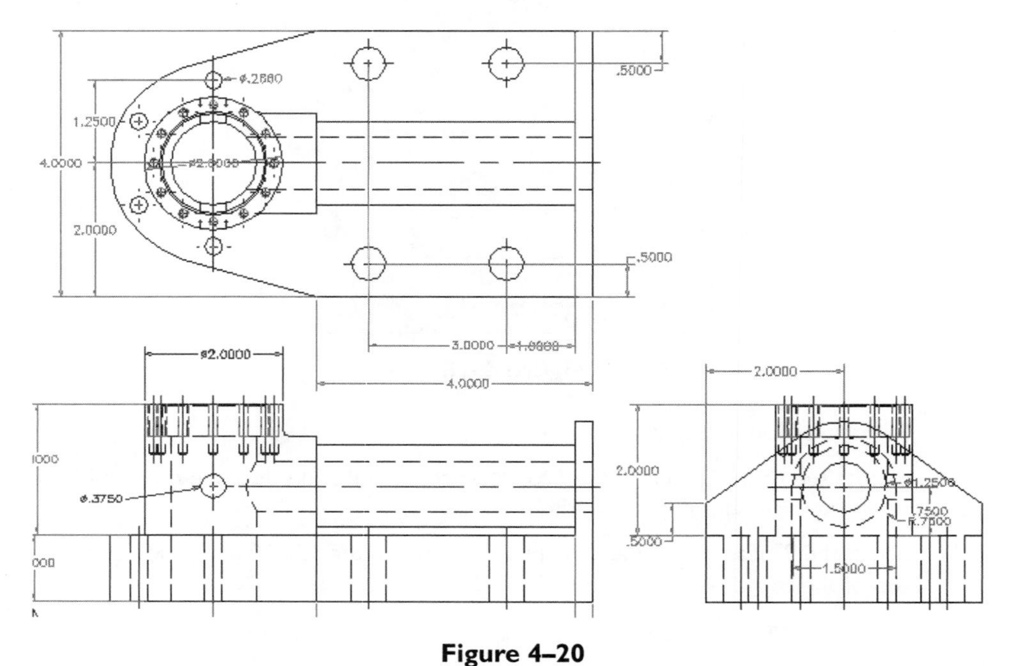

Figure 4–20

16. Turn on the AM_VIEWS layer.

17. Grip edit the newly created Ortho view to resemble Figure 4–21

Note: Parametric dimension and centerlines have been turned off in Figure 4–21 for clarity.

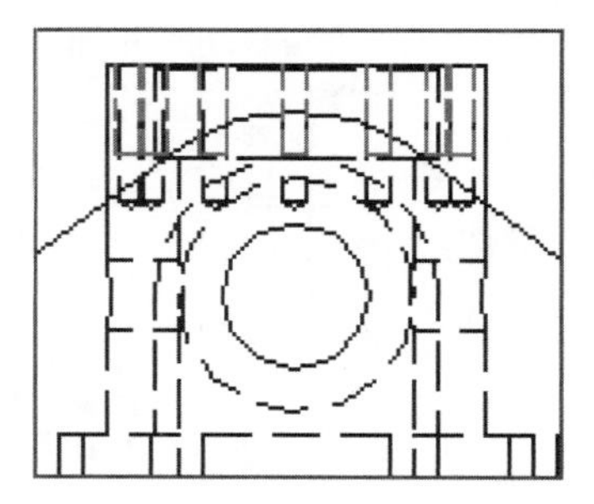

Figure 4–21

18. Zoom extents and then right click on the base view in the Browser. Select New View from the menu.

19. In the Create Drawing View dialog box, select Detail as the View Type, and enter a value of 2 for the Scale.

20. When prompted, select a vertex point in the base view.

21. When prompted for the center point, right click and select Polygon.

22. Sketch a polygon around the 2" diameter feature.

23. Place the view to the left of the base view. Your drawing should resemble Figure 4–22.

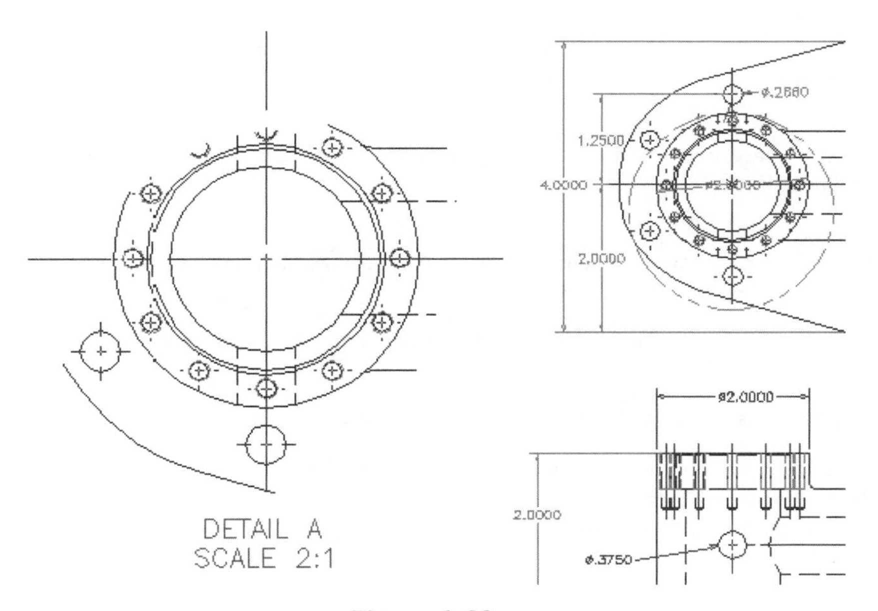

Figure 4–22

24. Right click on Layout1 in the Browser and select Export All Views.

25. Enter a filename and from the Files of Type: list, choose: AutoCAD 2000 DXF (*.dxf). Be sure to note where the file is being placed.

26. Press the save button.

27. Close Mechanical Desktop and open vanilla AutoCAD.

28. DXFIN the file just created.

REVIEW QUESTIONS

1. Active parts can now be selected for drawing views after a scene has been created. T or F?

2. Can items appearing as splines in a drawing view be dimensioned?

3. What type of dimension must be selected in order to use Object snaps (OSNAPS)?

4. Detail views can be created with non-rectangular viewports. T or F?

5. To use the select option for creating a detail view, what type of shape must already exist?

6. Resized viewports will always go back to the original size when drawing views are updated. T or F?

7. Drawing Views can only be exported to AutoCAD 2000i format. T or F?

8. Only one dimension at a time can be moved from one view to another. T or F?

9. Power Dimensions in drawing views support all Object Snaps. T or F?

10. Section View and Detail View labels will remain in their edited state when drawing view updates are done. T or F?

Web Based Enhancements

Several features have been added to Mechanical Desktop that allow you to use the Internet as a design tool. Immediately after starting Mechanical Desktop 5 a new tool called the Mechanical Desktop Today page greets you. From this window you can open new or existing files, use template files or startup wizards, as well as access basic and advanced design information from the Internet. As with seemingly all Internet sites, the only real constant is change. This chapter will be a very general overview of some of the core features of the new web-based functionality.

After completing this chapter, you will be able to:

- Navigate Within The MCAD Today Window
- Find And Determine The Need For Software Updates
- Find Discussion Groups

NAVIGATE THE MCAD TODAY WINDOW

The Mechanical Desktop Today window is separate from Mechanical Desktop and serves as a portal to various Internet functions. At the top of the window is the My Workplace area (see Figure 5–1). The My Drawings section contains three tabs for performing operations such as opening existing drawings, starting new drawings and using template files. To make it easier to access files, the Open Drawings tab contains a history of recently opened files. Selecting a file from the list automatically opens it. The history listing can be sorted by date, filename or location. If the file you want to open is not on the history list, you can use the Browse… link which calls up the familiar Select File dialog box.

On the right side of the My Workplace area is the Bulletin Board (see Figure 5–2). The Bulletin Board is an area where CAD managers can set up company standards, notify their user base of the changes, schedule meetings, and more. By selecting the Edit button in the upper right corner, you can specify a server to connect to, allowing you to perform the tasks listed above.

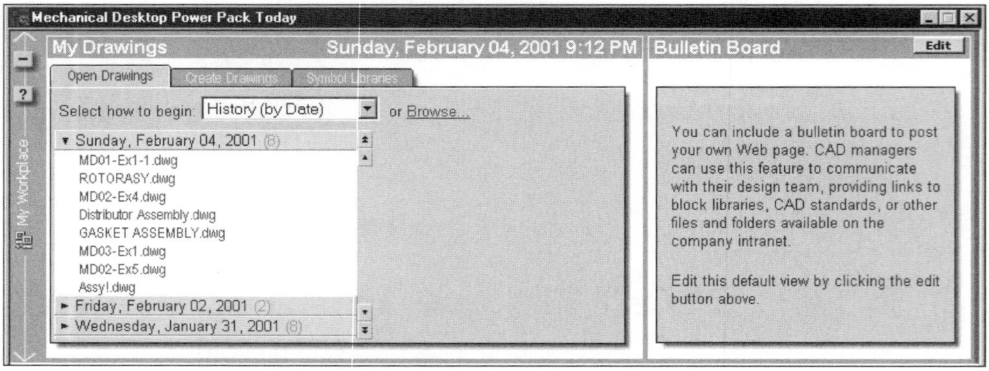

Figure 5–1

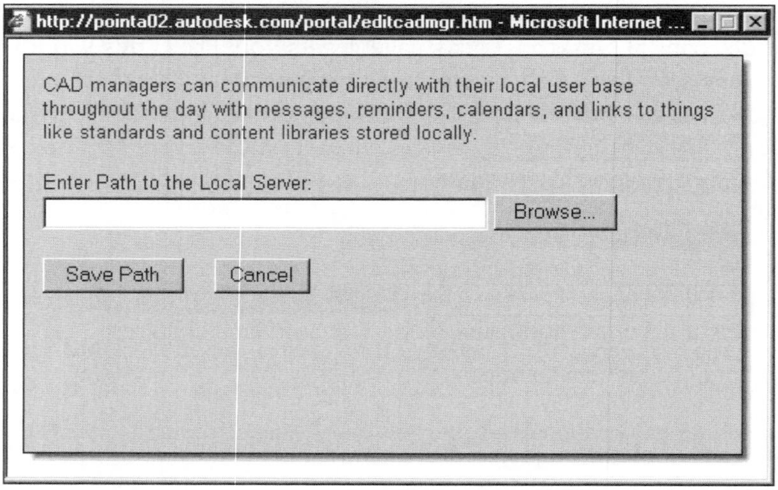

Figure 5–2

WEB BASED FUNCTIONS

The bottom of the Mechanical Desktop Today window holds the Internet functionality. This section is called The Web (see Figure 5–3). At the top of the Web area is Point A. Point A is the connection site, maintained by Autodesk, for all of the Internet utilities. When you select any of the category buttons to the right of the Point A link, specific information for that category is displayed. For instance, after selecting the News & Views category, the Web area changes to display a list of technology categories (see Figure 5–4). Depending on the technology selected, current headlines of interest regarding that aspect of technology are displayed in the middle column. The column to the far right displays current stock market indicators and allows you to check on specific stock information.

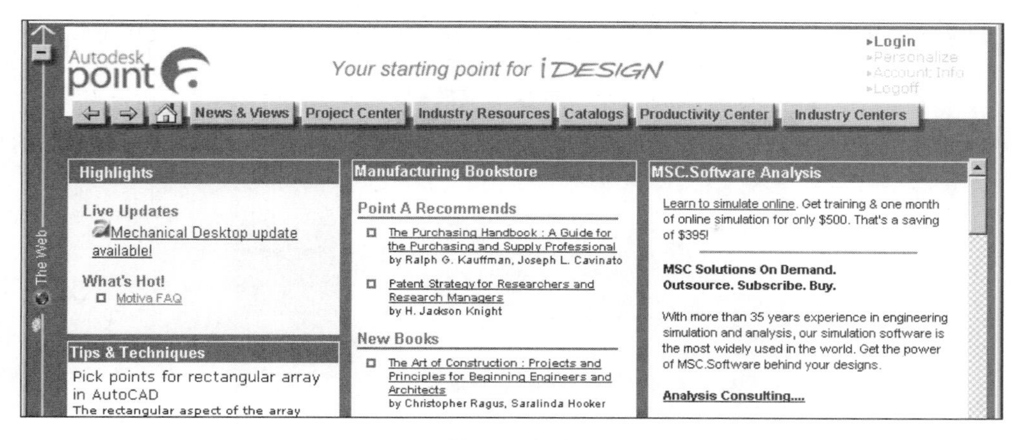

Figure 5–3

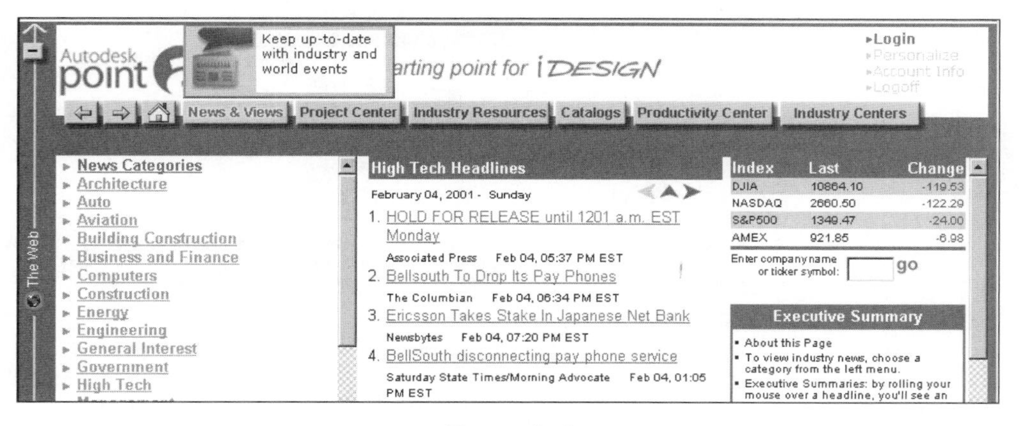

Figure 5–4

The Project Center button enables you to host projects on the web. This allows partners and customers to interact with designers. The Industry Resources category button provides access to items such as CAD related online magazines, a Units Converter, and other useful links. The Catalogs button connects you to a world of online symbol catalogs. You can search through catalogs of free symbols or those offered by various manufacturers. The Productivity Center button links you to tips and tricks links, discussion groups for worldwide support, e-learning classes, and a bookstore devoted to CAD industry.

FIND SOFTWARE UPDATES

When you first enter the Point A home page, one of the most important areas to note is the Highlights section. As shown in Figure 5–5, the Highlights area notifies you if any updates are available for Mechanical Desktop. When you select the Mechanical

Desktop Update available link!, a new page displays and offers instructions and requirements on downloading and installing the update (see Figure 5–6).

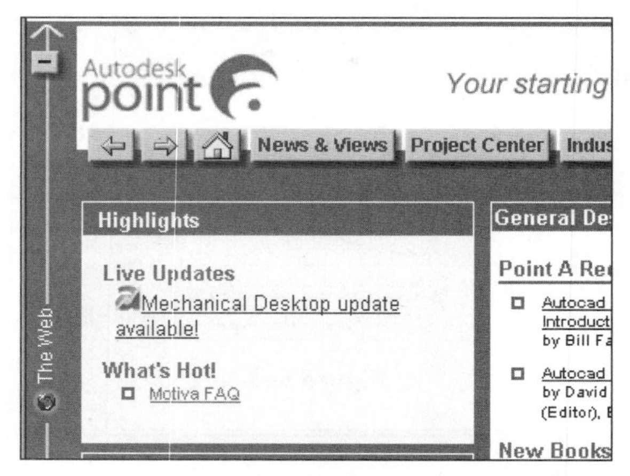

Figure 5–5

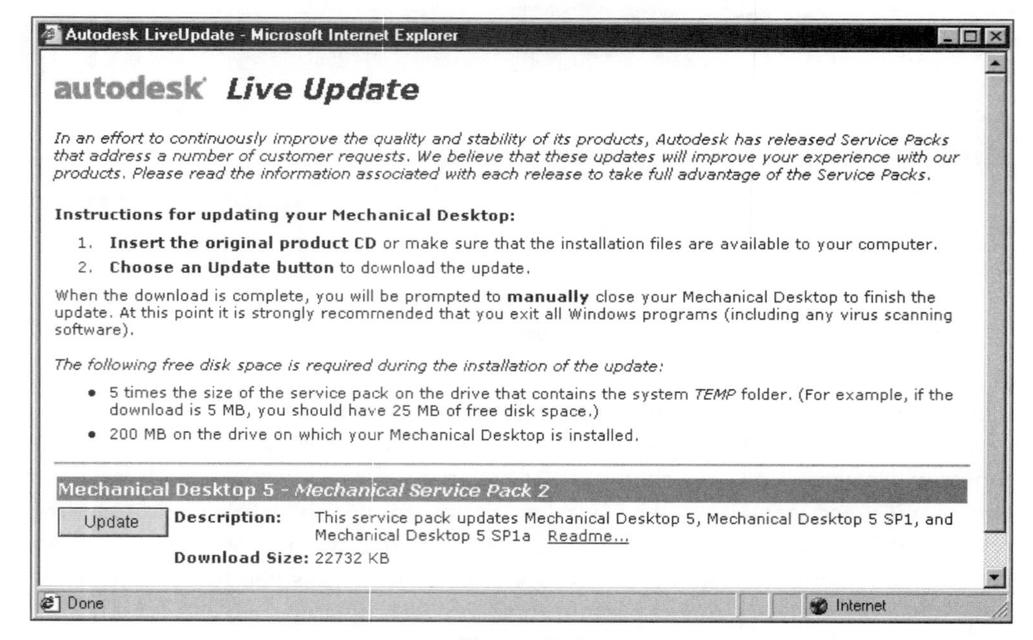

Figure 5–6

FIND DISCUSSION GROUPS

The ability to participate in discussion groups allows you to connect with other Mechanical Desktop users around the world, as well as Autodesk personnel. Discussion groups are appropriate places to ask "how-to" questions or solicit opinions on "why did this happen?". You can even post your files for others to look at and make recommendations. Figure 5–7 shows how the Product Discussion Groups item is accessed from the Productivity Center category button. Figure 5–8 shows a partial list of the discussion groups available. Virtually all Autodesk products have their own discussion groups. These groups use your current Internet browser to handle viewing and posting of messages.

Figure 5–7

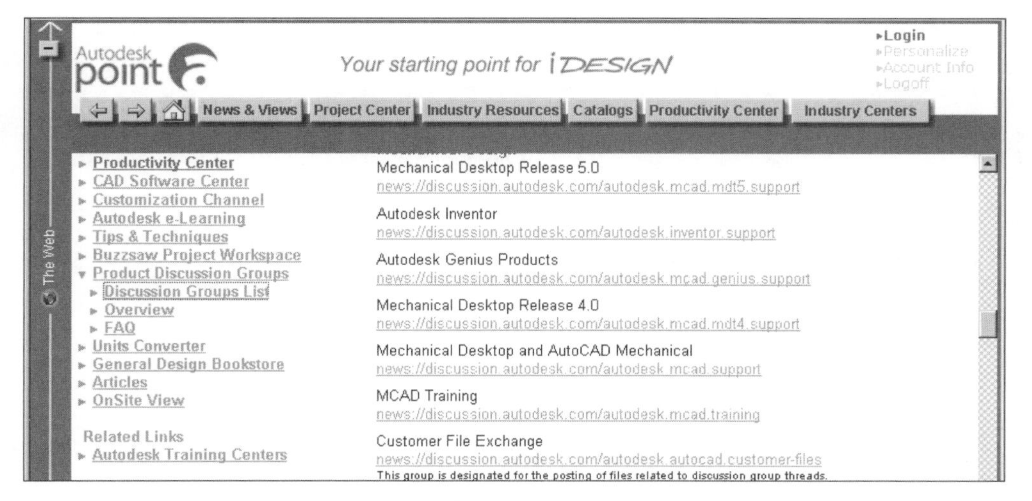

Figure 5–8

The preceding information highlighted a few of the many options available through the Mechanical Desktop Today window. You can also have information sent to you via email in the form of a newsletter, directly from the Point A site. As with any Internet site, change is constant and comes quickly. The information presented here reflected the state of the Mechanical Desktop Today window at the time of writing this book.

REVIEW QUESTIONS

1. You can open existing files, create new files, and use templates from the Mechanical Desktop Today window. T or F?

2. In the History list, files can be listed by date, filename and location. T or F?

3. Software updates can be accessed from the Mechanical Desktop Today window. T or F?

4. Discussion groups allow for worldwide support. T or F?

5. Tips and Techniques, Manufacturer's Catalogs, and a Units Converter are all items available in the Productivity Center. T or F?

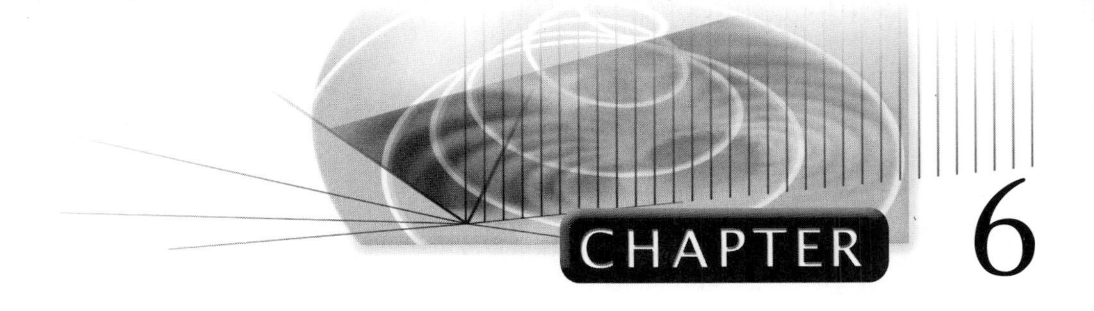

Introduction to the Power Pack

The Power Pack portion of Mechanical Desktop 5 is a vast collection of both 2D and 3D *"object oriented"* tools that automate and simplify the design process. This chapter concentrates on the 3D aspects of the Power Pack. The most obvious feature of the Power Pack is the Content 3D pull-down menu. However, portions of the Power Pack are prevalent throughout Mechanical Desktop 5. The Mechanical Main toolbar contains several "power" commands. Creating a Bill of Materials and Drawing Borders are just a few of the many time saving capabilities included in the Power Pack.

Unlike many other utilities, the Power Pack is not just a group of files or blocks that are installed and then inserted when needed. The Power Pack is actually a database of intelligent information that will build the items as they are specified. Therefore, depending on the specific part or feature, the type, size, or location of the item is often defined before running the actual command.

To comprehensively cover every aspect of the Power Pack would require a book of its own. This chapter will introduce you to some of the main features and capabilities of the Power Pack and how to use those features in your designs.

After completing this chapter, you will understand:

- Creating Holes
- Fasteners
- Screw Connections
- Shaft Generator
- Steel Shapes
- Power Commands: Dimension, Edit, Copy, Erase, Snap

CREATING 3D HOLES

If you are familiar with creating holes using the AMHOLE command, then you already understand some of the functionality of using the Power Pack to create holes. However, there are also many differences. The first difference is you have to specify what type of hole you want to create before you actually enter the command.

Figure 6–1 shows eight different types of holes that can be added to a Mechanical Desktop part, if you include slots as holes. Each option has its own dialog box where the specifications of the hole will be entered. For example, to place a simple ½" diameter, user-defined blind hole in the center of a 4" square part, select Blind Holes... from either the Content 3D pull-down menu or the Content 3D toolbar. The Select a Blind Hole dialog box appears, as shown in Figure 6–2.

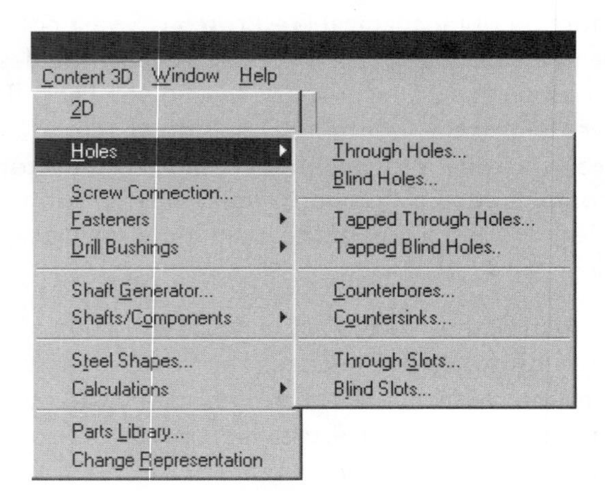

Figure 6–1

From this dialog box, select the type of hole to be created. The basic options are: Blind Hole, Blind Hole Metric, and User Blind Holes. The first two options contain all the standard size holes that are normally created. User-defined holes are used for all other applications. Since this example started out using English units, select the Blind Hole option. Next determine the type of placement for the hole. For this step, the Hole Position Method First Hole dialog box appears (see Figure 6–3).

As you can see from Figure 6–3, there are more options available for hole placement than when using the AMHOLE command. For this example, select the 2 Edges option. Using this option is exactly the same as using the 2 Edges option of the AMHOLE command. Once the location is determined, and the parametric dimension values are entered, the Blind Hole - Nominal Diameter dialog box appears, as shown in Figure 6–4.

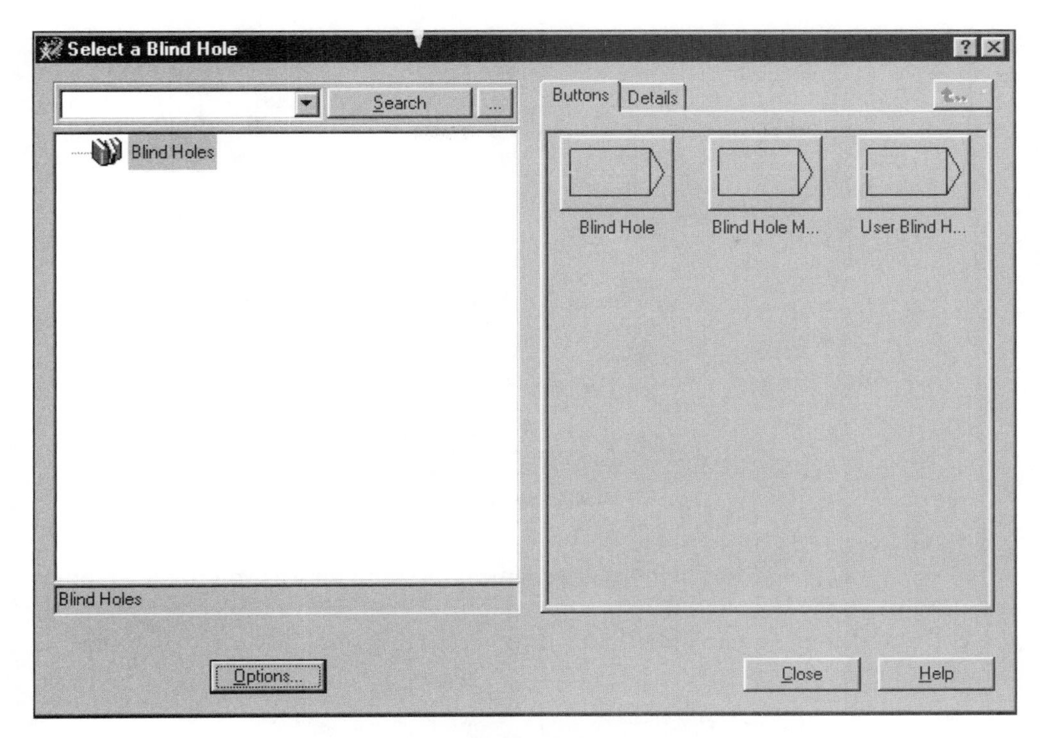

Figure 6–2

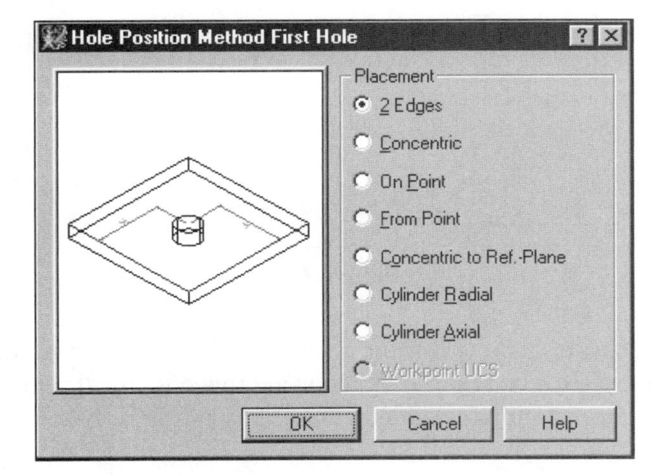

Figure 6–3

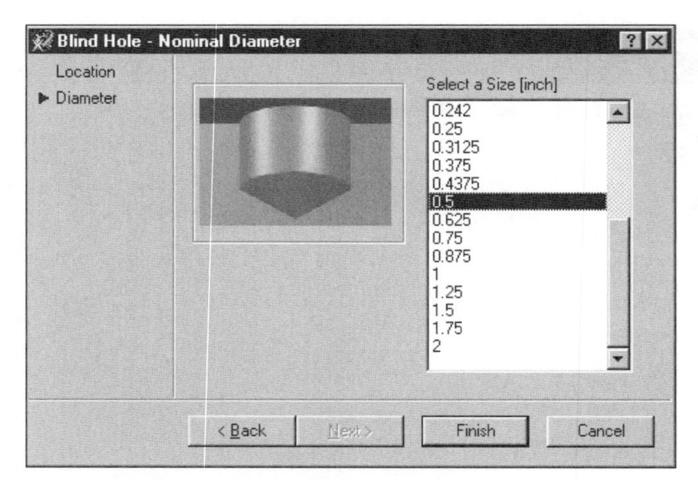

Figure 6–4

After selecting the size for the hole, Mechanical Desktop returns you to the graphics screen to place the hole and determine its depth. A numeric value can be entered at the command line, or enter the letter D to access the Enter Values dialog box, as shown in Figure 6–5.

Enter Values

Description	Variable	Minimum	Value	Maximum
Length:	LOTI	0.001	3.80184	

OK Cancel

Figure 6–5

If a nominal value is not used, select the User Blind Holes option to manually define the hole's diameter (see Figure 6–2). When using this option, after locating the hole, the User Blind Hole - Nominal Diameter dialog box appears, as shown in Figure 6–6.

Just as with any other features, once a hole is placed in the part, that feature will also appear in the Browser. However, its designation in the Browser differs from holes placed using the AMHOLE command. This unique designation makes it easy to determine what type of hole was placed, unlike the AMHOLE command. Figure 6–7 shows several different holes placed after an extrusion. Those designated Hole1 and Hole2 were placed using the AMHOLE command, but you cannot tell directly from the Browser that one is a drill-through hole and the other is a counterbored hole. The Power Pack makes it easy to distinguish the type of holes placed.

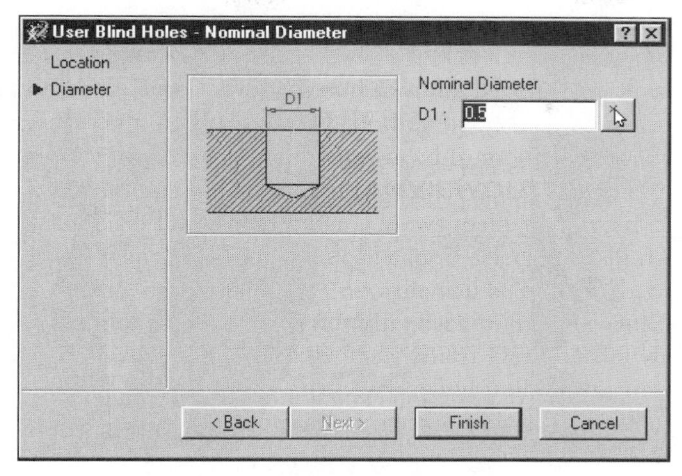

Figure 6–6

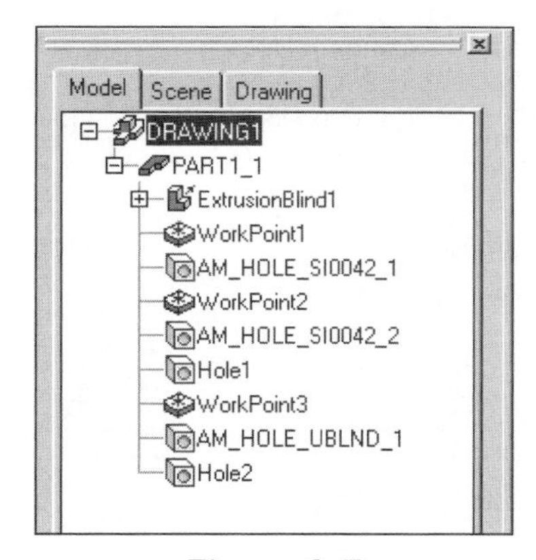

Figure 6–7

The procedure just outlined is common for all types of holes placed. However, each type of hole requires specific information to complete the feature. For example, a tapped hole requires that the tapped hole type be defined. Figure 6–8 shows the different options available for tapped through holes. After selecting the type and locating the tapped hole, Mechanical Desktop displays the appropriate dialog box so that the size can be determined. Figure 6–9 shows the options available for UNC threads. Mechanical Desktop sets the drill diameter based on the Standards selected when the program was installed.

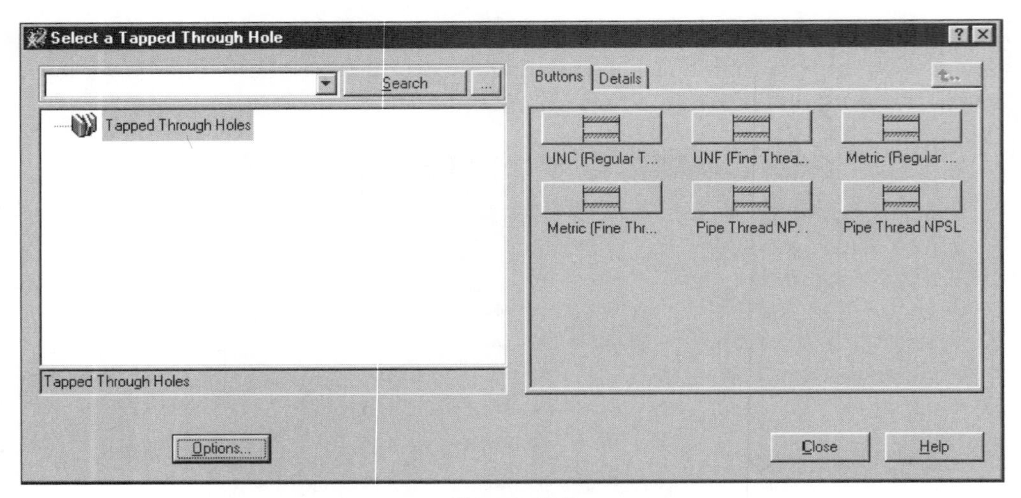

Figure 6–8

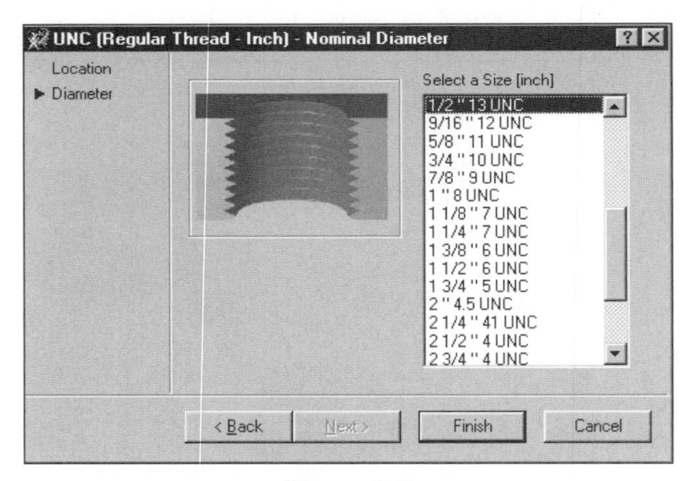

Figure 6–9

Editing a hole placed with the Power Pack is the same as editing any other hole in Mechanical Desktop. Use either the AMEDITFEAT command and select the hole, or right-click on the hole in the Browser to bring up the standard AMHOLE dialog box. Now values specific to the type of hole placed can be edited.

FASTENERS

Fasteners, as the name implies, allow you to create many different types of bolts, screws, nuts, washers, and other fasteners quickly. Figure 6–10 shows the list of available fasteners from the Content 3D menu. The placement of these fasteners is similar to assigning different types of holes, specific information is needed for each application. Figures 6–11 through 6–13 show the Select a Screw, Select a Washer and Select a Nut dialog boxes. These are the more common dialog boxes used when creating fasteners with the Mechanical Desktop Power Pack.

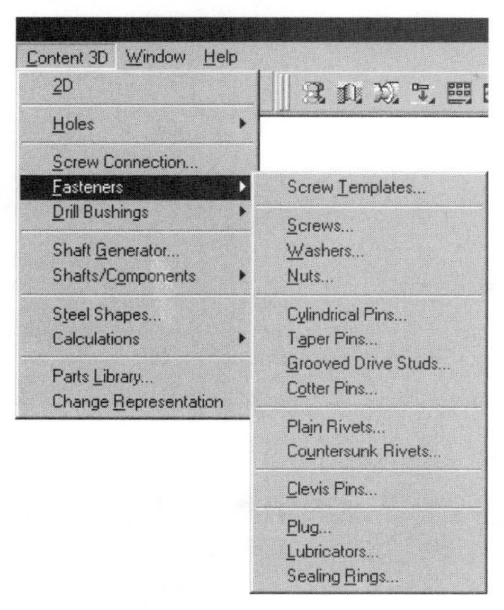

Figure 6–10

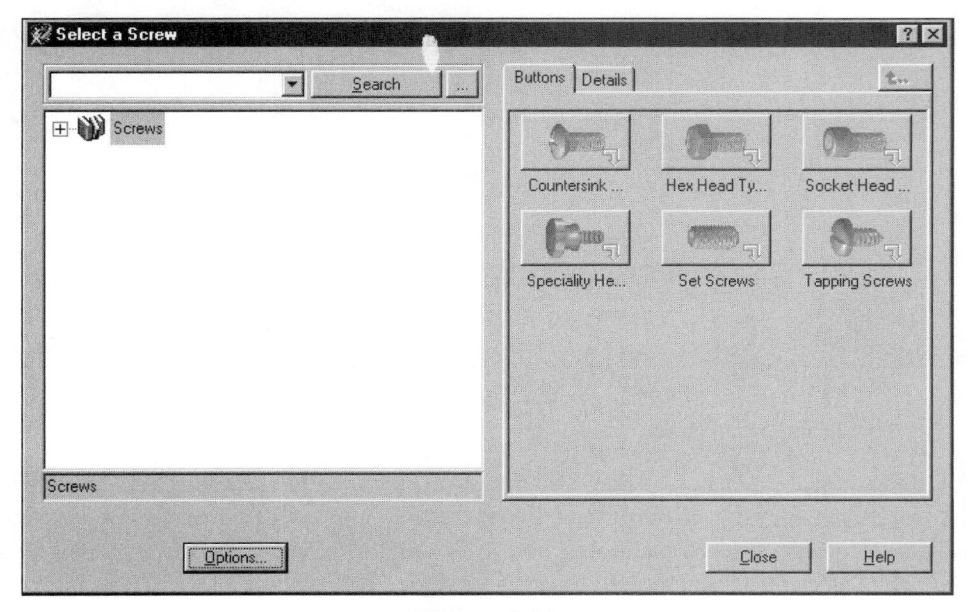

Figure 6–11

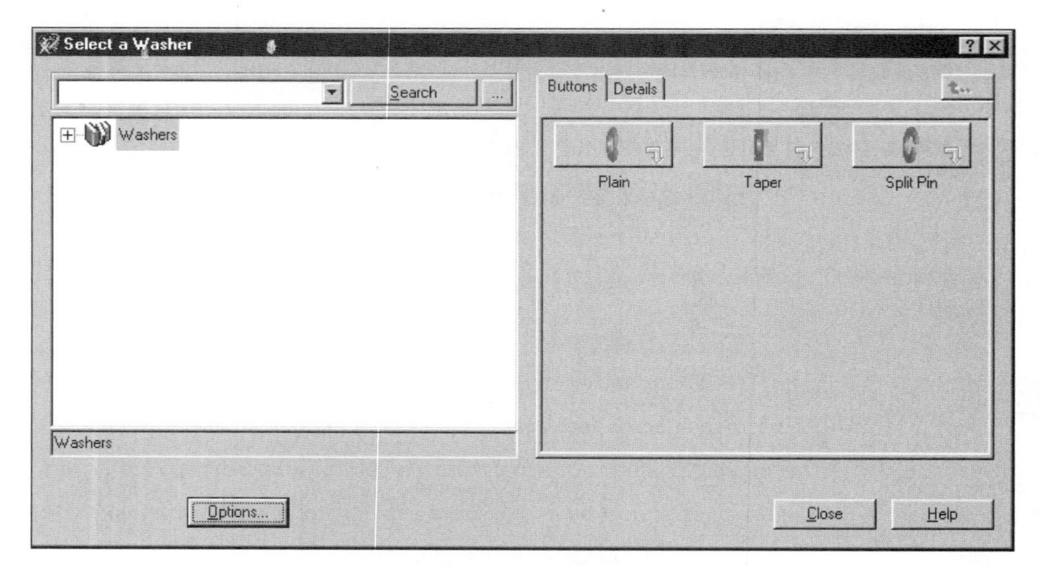

Figure 6–12

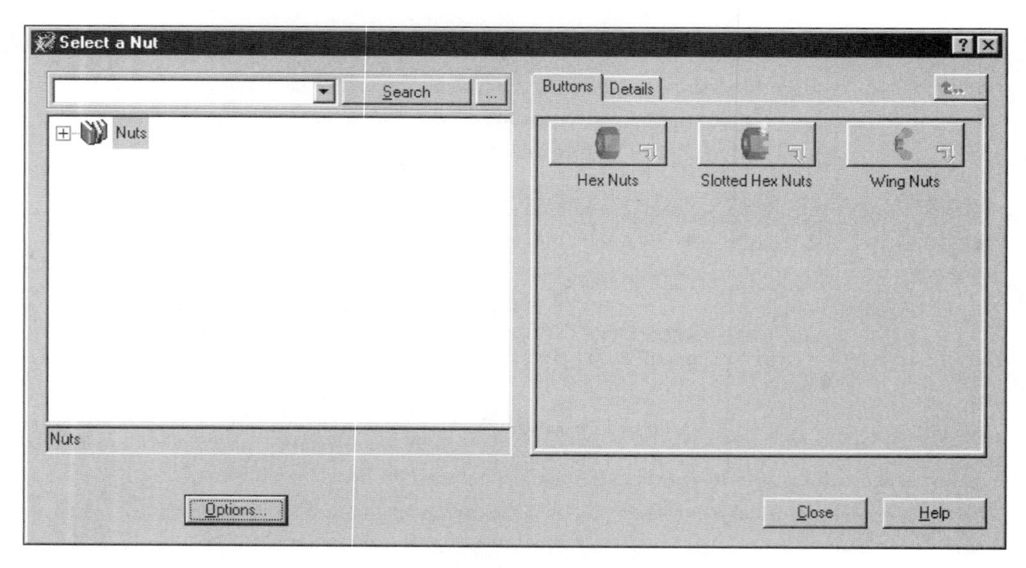

Figure 6–13

For this example, place a Socket Head Cap Screw into the drawing. As shown in Figure 6–11, select the Socket Head Types button. You could also expand the menu listing on the left side of the dialog box to locate and select the screw type. At this level, the selection is very general in nature. Once the type is selected, the buttons portion of the dialog box changes to show the specific types available. Scroll down, through the over 70 different screws available, to find the Socket Head Cap Screw, shown in Figure 6–14.

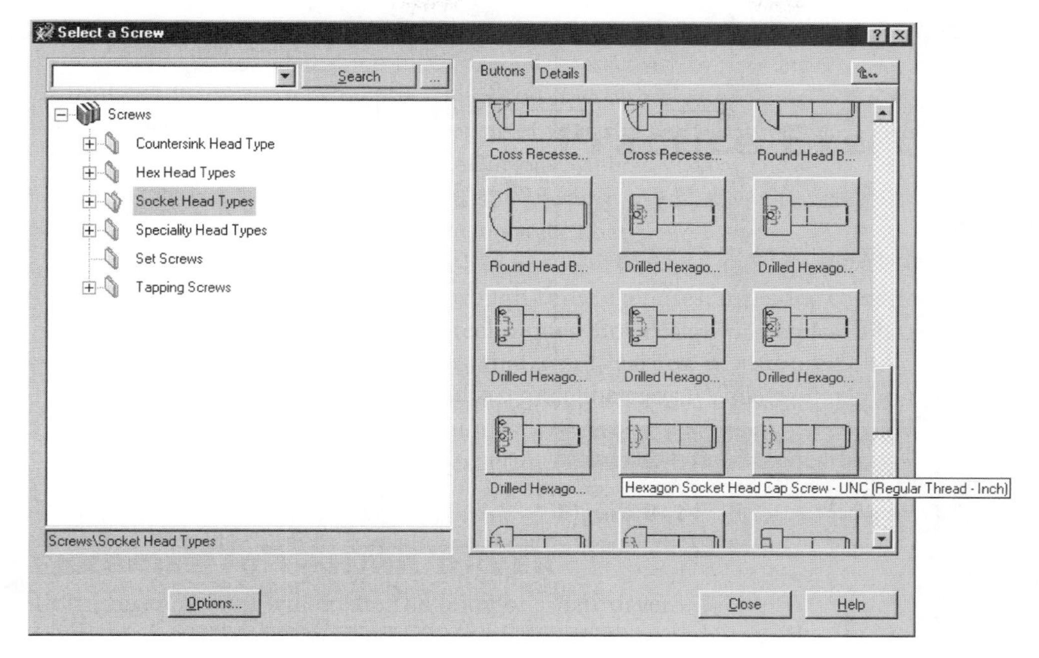

Figure 6–14

The next step is to place the screw into the drawing. At the command line, Mechanical Desktop gives you several options for placement (see Figure 6–15). The fastener you are placing is going to be treated as a new part in the drawing file and appears as such in the Browser (see Figure 6–16). Mechanical Desktop gives you the option to place the screw exactly where it will be used in the assembly. While these placement options may appear similar to those used to parametrically placed hole features, they do not control the fastener after it has been placed. These placement options are *not* the same as applying assembly constraints. The newly added fasteners will not move along if the base part is moved. Regardless of the method used to place the fastener, if there is a chance the part used to define the location of the fastener, or the fastener itself, will move, assembly constraints must be used to maintain the proper location of the fasteners.

The initial option for placement is to select a first point. Mechanical Desktop then prompts you to select a second point. The two points selected define the centerline, or axis of the fastener The length of the fastener is defined later. The Concentric and two Edges options work the same way with fasteners as they do when placing a hole using the AMHOLE command. The Cylinder option allows you to select a cylindrical face but use the cursor to locate it on that face.

```
Command:
Command: _amscrew3d
Select first point [Concentric/cYlinder/two Edges]:
Target: DRAWING1 5.1163, -0.6471, 0.0000    SNAP GRID ORTHO POLAR OSNAP OTRACK LWT MODEL
```

Figure 6–15

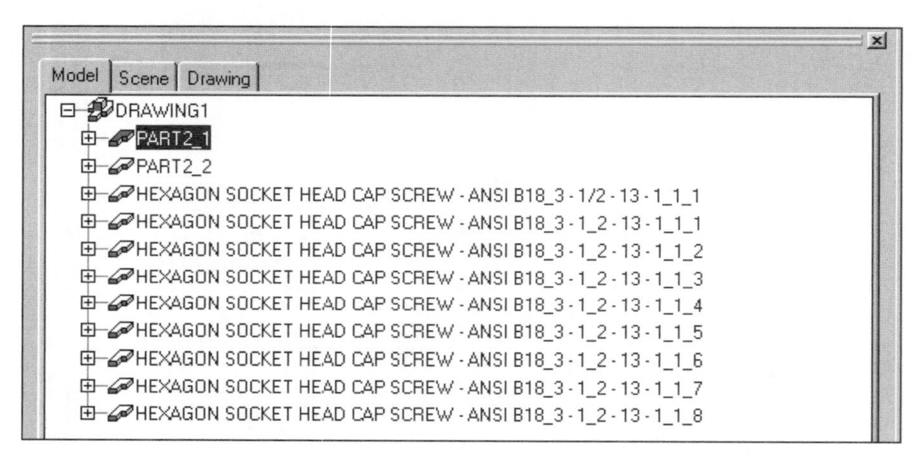

Figure 6–16

Once the fastener has been located, the length can be determined to complete the part. This can be done in three ways. First, a numerical value for the length can be entered at the command line. Second, you can drag the cursor and Mechanical Desktop will update the fastener length based on the standard sizes of the fastener. Third, you can enter the letter D at the command line, as shown in Figure 6–17, and a dialog box containing all the standard sizes available for the screw selected appears. You can pick from this list and view the other data pertinent to the standard part selected, as shown in Figure 6–18. The entry highlighted in Figure 6–18 is for the Socket Head Cap Screw that you will be using.

Figure 6–19 shows a proximity switch and a mounting bracket that need to be bolted together. The diameter of the holes has been increased for clarity. Using the option for socket head cap screws, select 3/8 UNF to fit the holes in the back of the switch. Once located, go to the dialog box to determine the length from the standard sizes available. In Figure 6–20 you can see the completed assembly. You can also see that some washers were used under the socket head cap screws. That will be addressed in the next section.

```
Select circular or elliptical edge:
Choose insertion direction [Flip/Accept] <Accept>:
Drag size [Dialog]:
```

> ANSI B18.3 - 1/2 - 13 - 1/2 < -0.5163, 0.7969 , 0.0000 SNAP GRID ORTHO POLAR OSNAP OTRACK LWT MODEL

Figure 6–17

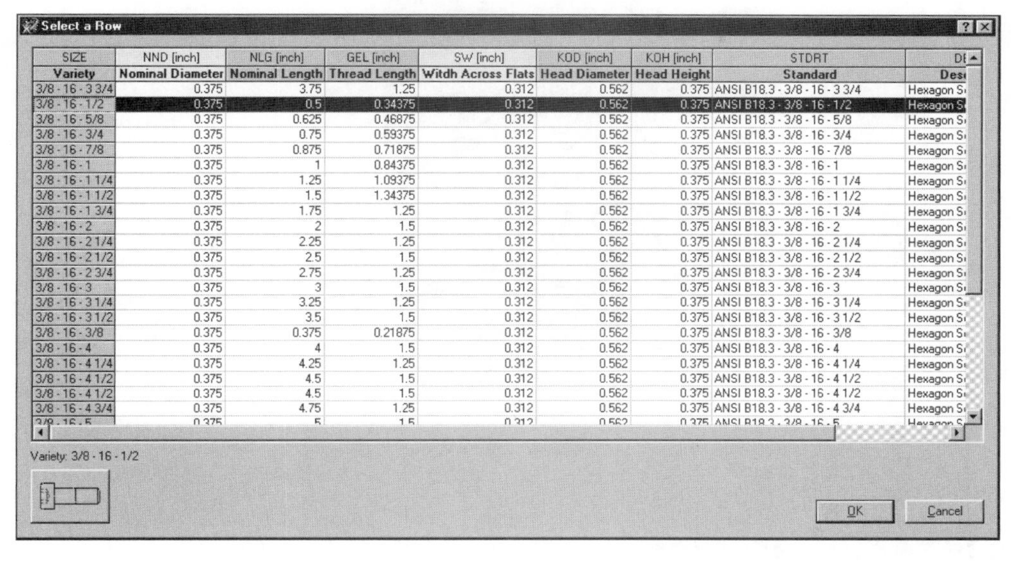

Select a Row

SIZE	NND [inch]	NLG [inch]	GEL [inch]	SW [inch]	KOD [inch]	KOH [inch]	STDRT	DE
Variety	**Nominal Diameter**	**Nominal Length**	**Thread Length**	**Witdh Across Flats**	**Head Diameter**	**Head Height**	**Standard**	**Des**
3/8 - 16 - 3 3/4	0.375	3.75	1.25	0.312	0.562	0.375	ANSI B18.3 - 3/8 - 16 - 3 3/4	Hexagon S
3/8 - 16 - 1/2	0.375	0.5	0.34375	0.312	0.562	0.375	ANSI B18.3 - 3/8 - 16 - 1/2	Hexagon S
3/8 - 16 - 5/8	0.375	0.625	0.46875	0.312	0.562	0.375	ANSI B18.3 - 3/8 - 16 - 5/8	Hexagon S
3/8 - 16 - 3/4	0.375	0.75	0.59375	0.312	0.562	0.375	ANSI B18.3 - 3/8 - 16 - 3/4	Hexagon S
3/8 - 16 - 7/8	0.375	0.875	0.71875	0.312	0.562	0.375	ANSI B18.3 - 3/8 - 16 - 7/8	Hexagon S
3/8 - 16 - 1	0.375	1	0.84375	0.312	0.562	0.375	ANSI B18.3 - 3/8 - 16 - 1	Hexagon S
3/8 - 16 - 1 1/4	0.375	1.25	1.09375	0.312	0.562	0.375	ANSI B18.3 - 3/8 - 16 - 1 1/4	Hexagon S
3/8 - 16 - 1 1/2	0.375	1.5	1.34375	0.312	0.562	0.375	ANSI B18.3 - 3/8 - 16 - 1 1/2	Hexagon S
3/8 - 16 - 1 3/4	0.375	1.75	1.25	0.312	0.562	0.375	ANSI B18.3 - 3/8 - 16 - 1 3/4	Hexagon S
3/8 - 16 - 2	0.375	2	1.5	0.312	0.562	0.375	ANSI B18.3 - 3/8 - 16 - 2	Hexagon S
3/8 - 16 - 2 1/4	0.375	2.25	1.25	0.312	0.562	0.375	ANSI B18.3 - 3/8 - 16 - 2 1/4	Hexagon S
3/8 - 16 - 2 1/2	0.375	2.5	1.5	0.312	0.562	0.375	ANSI B18.3 - 3/8 - 16 - 2 1/2	Hexagon S
3/8 - 16 - 2 3/4	0.375	2.75	1.25	0.312	0.562	0.375	ANSI B18.3 - 3/8 - 16 - 2 3/4	Hexagon S
3/8 - 16 - 3	0.375	3	1.5	0.312	0.562	0.375	ANSI B18.3 - 3/8 - 16 - 3	Hexagon S
3/8 - 16 - 3 1/4	0.375	3.25	1.25	0.312	0.562	0.375	ANSI B18.3 - 3/8 - 16 - 3 1/4	Hexagon S
3/8 - 16 - 3 1/2	0.375	3.5	1.5	0.312	0.562	0.375	ANSI B18.3 - 3/8 - 16 - 3 1/2	Hexagon S
3/8 - 16 - 3/8	0.375	0.375	0.21875	0.312	0.562	0.375	ANSI B18.3 - 3/8 - 16 - 3/8	Hexagon S
3/8 - 16 - 4	0.375	4	1.5	0.312	0.562	0.375	ANSI B18.3 - 3/8 - 16 - 4	Hexagon S
3/8 - 16 - 4 1/4	0.375	4.25	1.25	0.312	0.562	0.375	ANSI B18.3 - 3/8 - 16 - 4 1/4	Hexagon S
3/8 - 16 - 4 1/2	0.375	4.5	1.5	0.312	0.562	0.375	ANSI B18.3 - 3/8 - 16 - 4 1/2	Hexagon S
3/8 - 16 - 4 1/2	0.375	4.5	1.5	0.312	0.562	0.375	ANSI B18.3 - 3/8 - 16 - 4 1/2	Hexagon S
3/8 - 16 - 4 3/4	0.375	4.75	1.25	0.312	0.562	0.375	ANSI B18.3 - 3/8 - 16 - 4 3/4	Hexagon S
3/8 - 16 - 5	0.375	5	1.5	0.312	0.562	0.375	ANSI B18.3 - 3/8 - 16 - 5	Hexagon S

Variety: 3/8 - 16 - 1/2

OK Cancel

Figure 6–18

Figure 6–19

Figure 6–20

SCREW CONNECTIONS

You could use fasteners to individually create each nut, bolt, and washer or, you could use the Screw Connection dialog box to build each part of the fastener in a single command. Selecting Screw Connection from the Content 3D pull-down menu or the Content 3D toolbar, brings up the Screw Connection dialog box, shown in Figure 6–21.

The Screw Connection dialog box is basically a combination of the various other fasteners. Usually the type of screw to be used is the first item defined in the dialog box, however, you can start with any component. Once the first component is defined, all subsequent items will default to the size of the first. In the case of the proximity switch, a single 3/8-16 SHCS was placed. You could go back and individually add washers to the screw, or you could use the Screw Connection dialog box to build the connection for you.

Start off by accessing the Screw Connection command and selecting the <Screws> button at the top of the list (see Figure 6–21). The Screw Connection dialog box takes you through the same process for selecting the screw as the fastener command did previously. Once the type of screw is selected, you are returned to the Screw Connection dialog box with the list of available sizes displayed on the right side of the dialog box. Figure 6–22 shows the selection of the 3/8-16 SHCS.

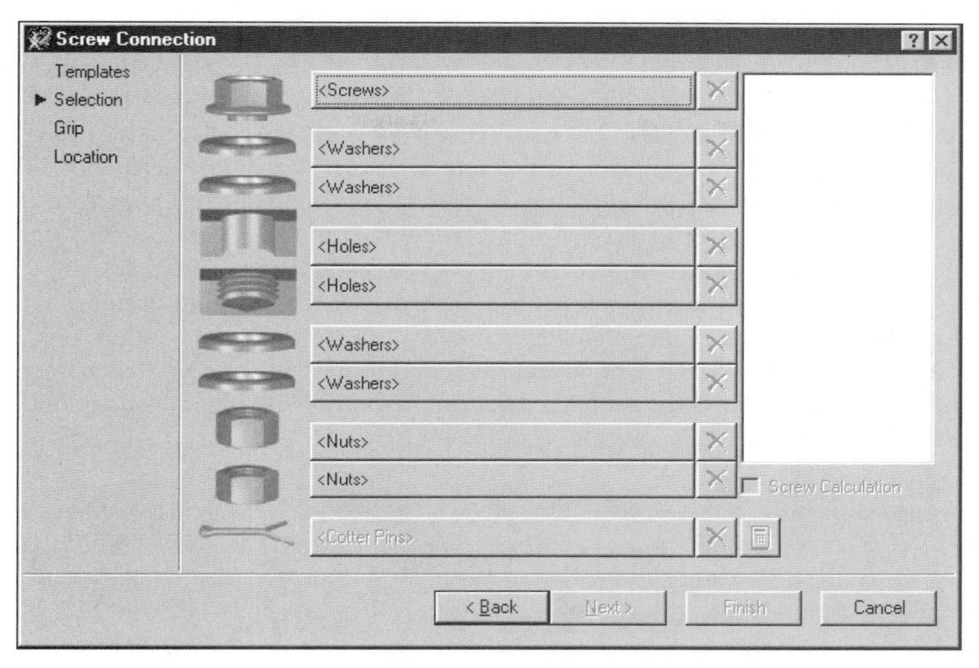

Figure 6–21

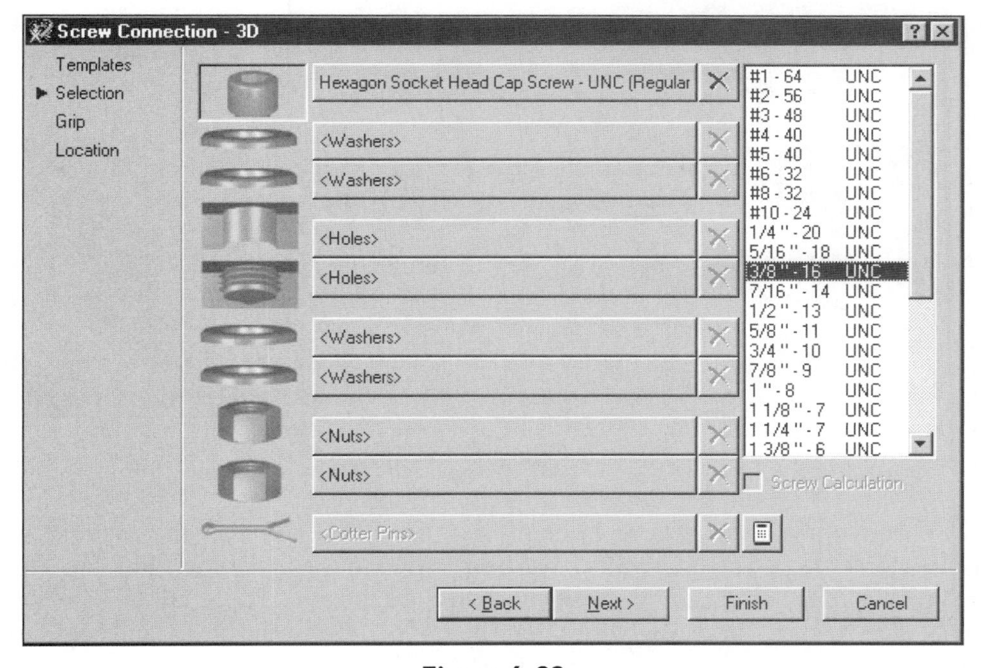

Figure 6–22

At this point you could select the next button and define the SHCS on screen just as you did before, or you could select another option from the Screw Connection dialog box to add more components. The next selection you will make is a washer. For this example, select a plain washer. After selecting the <Washer> button, a new screen appears listing the different types of washers available. Again, define the specific washer type here and then select the size needed by returning to the Screw Connection dialog box main screen. The dialog box will default to the same size as the screw selected, in this case that's 3/8" but you can change that size if needed. Also, as each component is defined, your selections are displayed on the selection buttons. This allows you to see what components have already been defined (see Figure 6–23). From here, place the Screw Connection just as you did with the previous components. Figure 6–24 shows the completed Screw Connection in place, verses the single fastener. The next example uses more components.

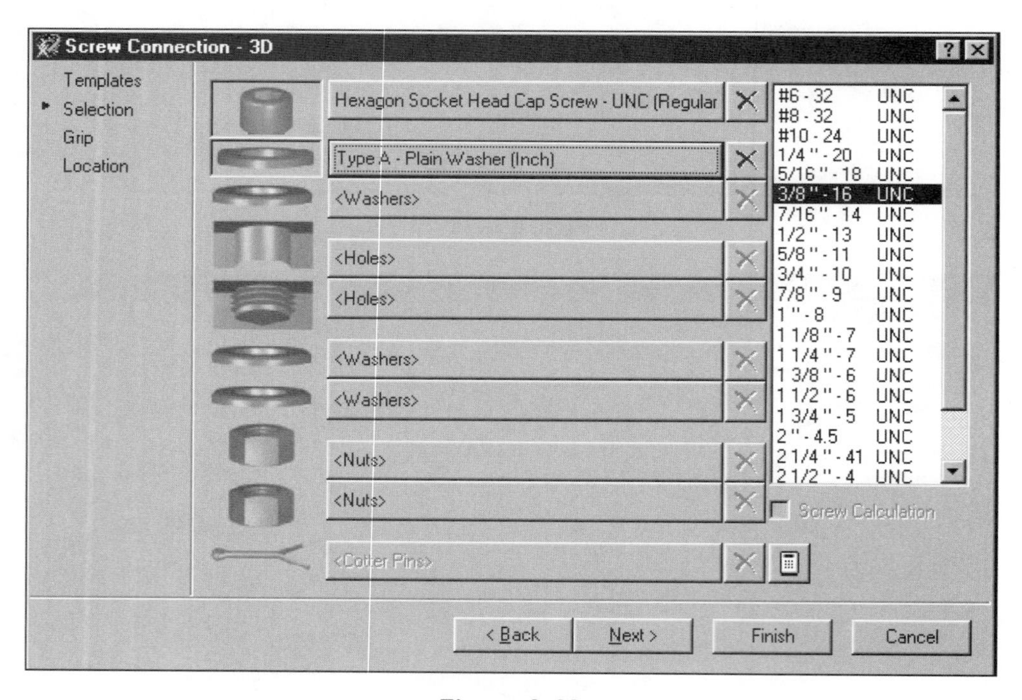

Figure 6–23

The next example uses a variety of options from the Screw Connection dialog box to secure a bracket. A Stand Off Block has been added to the back of the bracket. You are going to put a single screw connection through the bracket. Figure 6–25 shows the Stand Off Block in place. Notice that there is no hole in the block, and previous figures show no hole in the mounting plate.

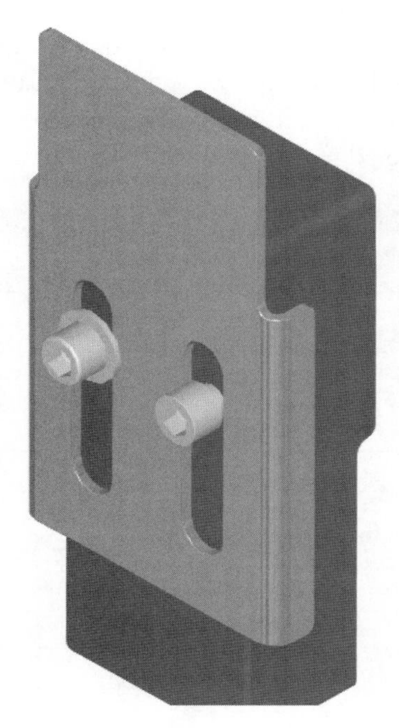

Figure 6–24

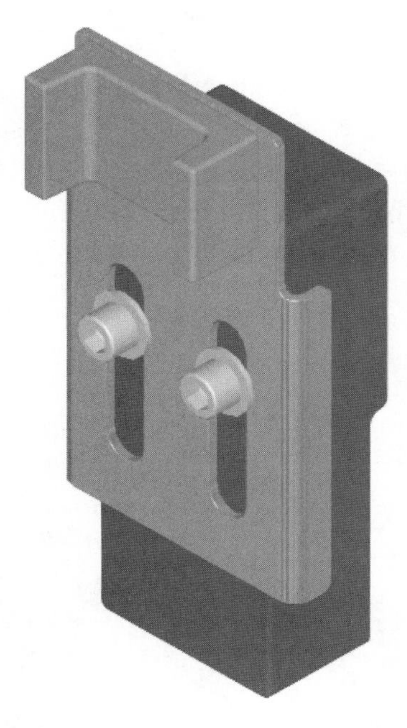

Figure 6–25

Figure 6–26 shows the Screw Connection dialog box complete with 3/8-16 SHCS, a Hi-Collar lock washer, two holes (one for the plate and one for the block), a plain washer, and a 3/8-16 hex nut. To aid in placing the screw connection, we have suppressed the fillet around the edge of the mounting plate.

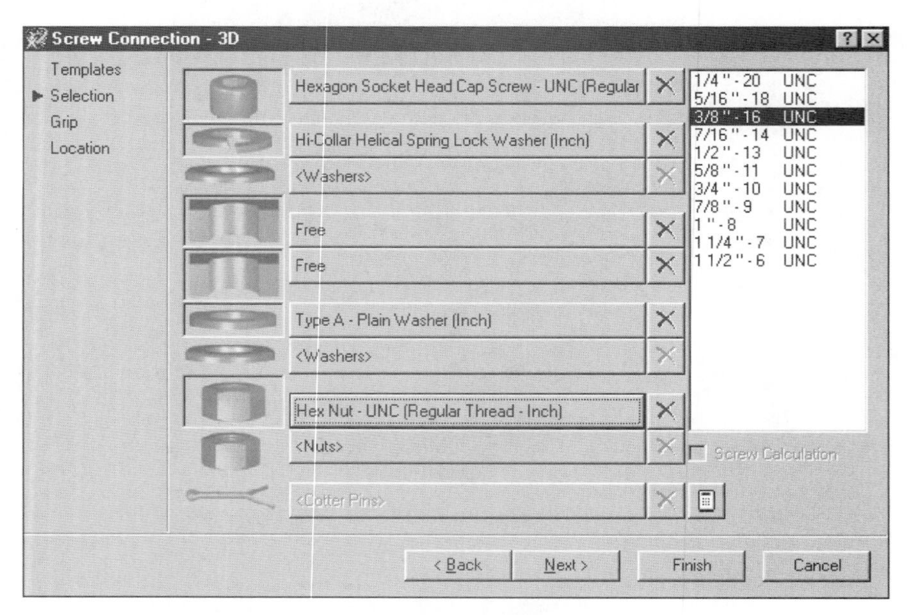

Figure 6–26

Figures 6–27 through 6–29 show the completed Screw Connection. In Figure 6–29, the Screw Connection has been suppressed to show that holes were created in both parts by using the Screw Connection.

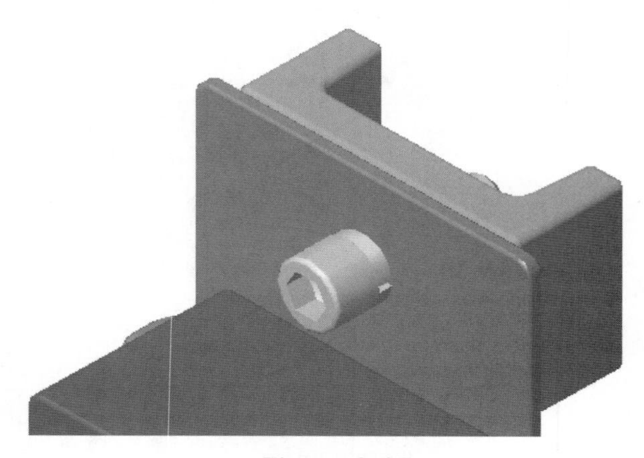

Figure 6–27

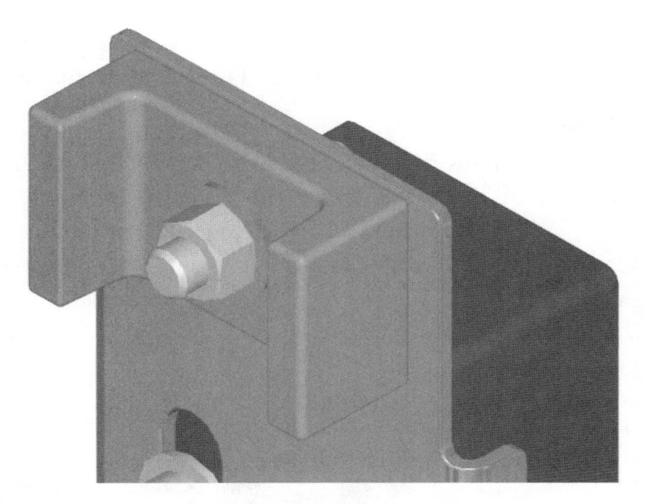

Figure 6–28

Figure 6–29

SHAFT GENERATOR

The Shaft Generator is another tool that greatly simplifies the design process. Like the Screw Connection, the building of the shaft actually takes place in a dialog box. The Mechanical Desktop Power Pack takes care of setting the work and sketch planes, creating the sketches, profiles, and finally, the features.

The first step in creating a new shaft is to define the centerline. Mechanical Desktop asks for a starting point to define the new shaft, or to pick an existing shaft for editing, when executing the AMSHAFT3D command. After the first point is established, you are prompted for the centerline endpoint. Next, you are prompted to select a third point to define the plane you will be working in. The default setting is parallel to the UCS.

As soon as the working plane is defined, a new part appears in the Browser and the 3D Shaft Generator dialog box appears, as shown in Figure 6–30. The 3D Shaft Generator dialog box is made up of several areas. The first area has three tabs used to define the Left Contour, Outer Contour, and Right Contour. When creating a new shaft, only the Outer Contour tab will be available. Once an Outer Contour is established, the Left Inner and Right Inner tabs become available. Each tab has a section on the right called the View Section. This section allows you to view the shaft from the front, side, or to rotate to any orientation while remaining in the 3D Shaft Generator dialog box.

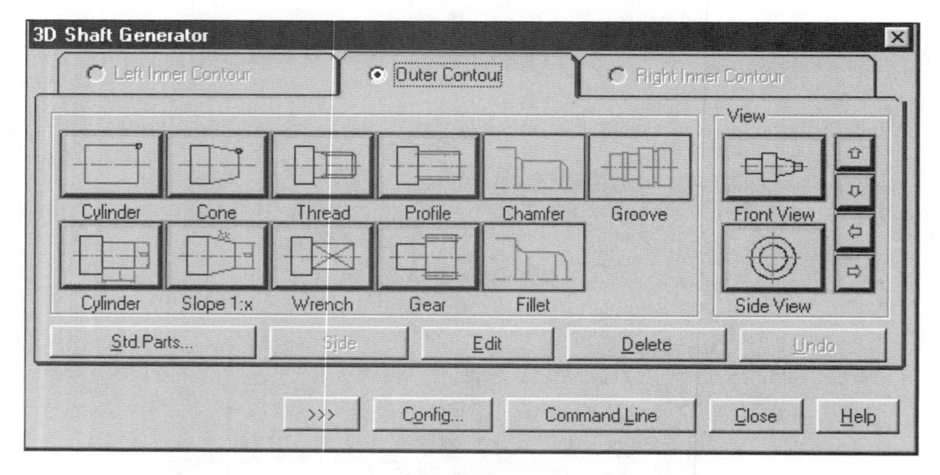

Figure 6–30

Below the input and view controls are buttons used for editing the shaft being built. The first button, labeled Std. Parts…, takes you to the Please Select a Part dialog box, shown in Figure 6–31. In this dialog box you select parts such as Roller Bearings, Circlips, Seals and more. The parts selected are based on the available parts defined by the current standard. The Side button controls which end of the shaft the next feature will be applied to. The Edit button allows you to go back and edit specific features of the shaft. At the bottom of the dialog box are the Config…, Command Line, Close, and Help buttons.

The Config… button sets the display of gears created. By default, only one full, and two half teeth of a gear are created. This is done to conserve system resources and to keep file size down. The Config… button allows you to display all the teeth, if desired. The Command Line button removes the dialog box so that all functions can be completed from the command line for that application of the 3D Shaft Generator. Reentering the command or using the DIalog option will return you to the dialog box. The Close button completes the shaft.

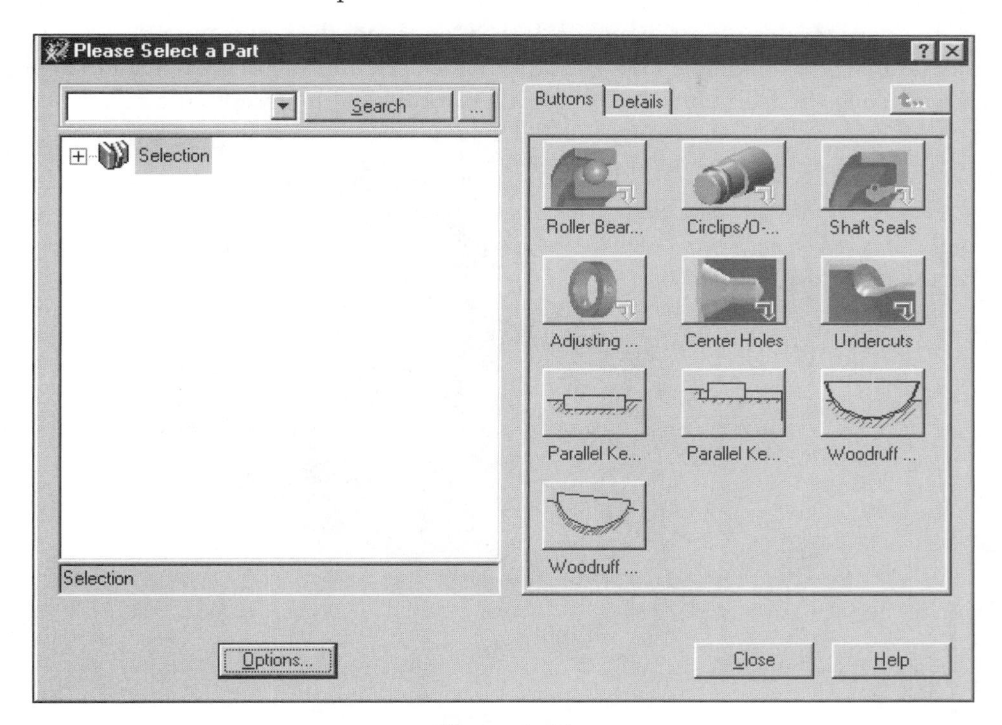

Figure 6–31

The Outer Contour of the shaft is built by selecting from the available buttons. The Cylinder button creates an extrusion feature. After selecting the Cylinder button, the 3D Generator dialog box disappears and the command line prompts you to: Specify other corner point. Think of creating a shaft as sketching the general shape. Based on your selection point, Mechanical Desktop will sketch a circle and extrude to the diameter and length specified on screen. An extrusion feature appears in the Browser along with any work features required to create and constrain the cylinder.

To set the exact size of the cylinder, use the Edit button at the bottom of the Outer Contour tab. Selecting the Edit button is similar to running the AMEDITFEAT command. You will be prompted to select the feature to be edited. Once the feature is specified, the appropriate dialog box appears along with the sketch. In the case of the cylinder, the Custom Feature: Extrusion dialog box appears, allowing you to set the distance of the extrusion. Parametric dimensions for the sketch also appear and can be edited by selecting the dimension to change. Once all changes have been made, the part updates and the 3D Shaft Generator dialog box reappears.

The Cone button creates a shape that has a linear change in diameter through the length of the feature. If the Cone is the first feature of the shaft, you are prompted for both the starting and ending diameters. If the Cone is not the first feature, the starting diameter matches the diameter of the last feature. The starting diameter can then be changed using the Edit button. A Cone is a revolved feature (see Figure 6–32).

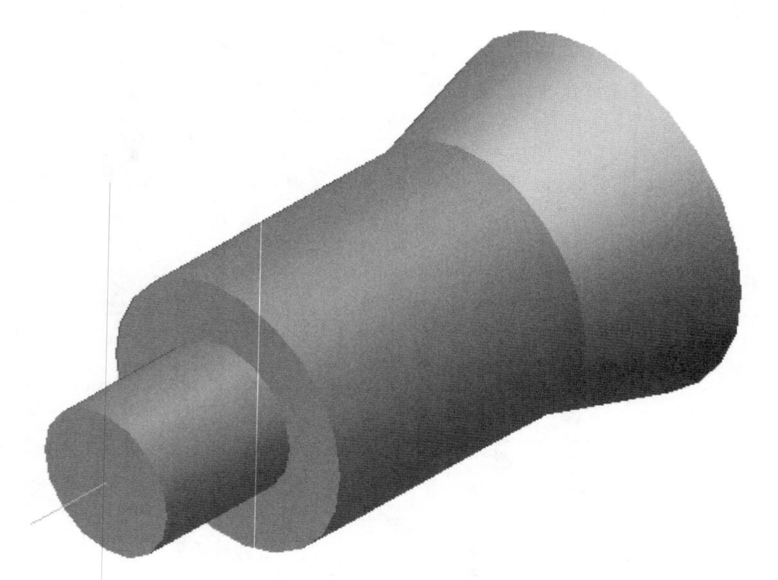

Figure 6–32

The Thread button takes you to the Thread dialog box. The number of standards selected during the installation process limits the thread options available. Figure 6–33 shows some of the available options. Once a thread type is specified, a dialog box appears allowing you to further define the thread to be created. Figure 6–34 shows the dialog box for an external ANSI thread. The desired size is selected from the standard size window. Undercuts can be added, and other features can be set using the image tile as a guide for each setting. Just as tapped holes don't appear when using the AMHOLE command, threads won't appear on the model but they will appear in the drawing views.

The Profile button is used to add splines to the shaft. Selecting the Profile button takes you to the Profile dialog box where you can select the type of spline to be created. Just like the Thread button, the splines available are based on the standards that have been installed. Figure 6–35 shows some of the available options.

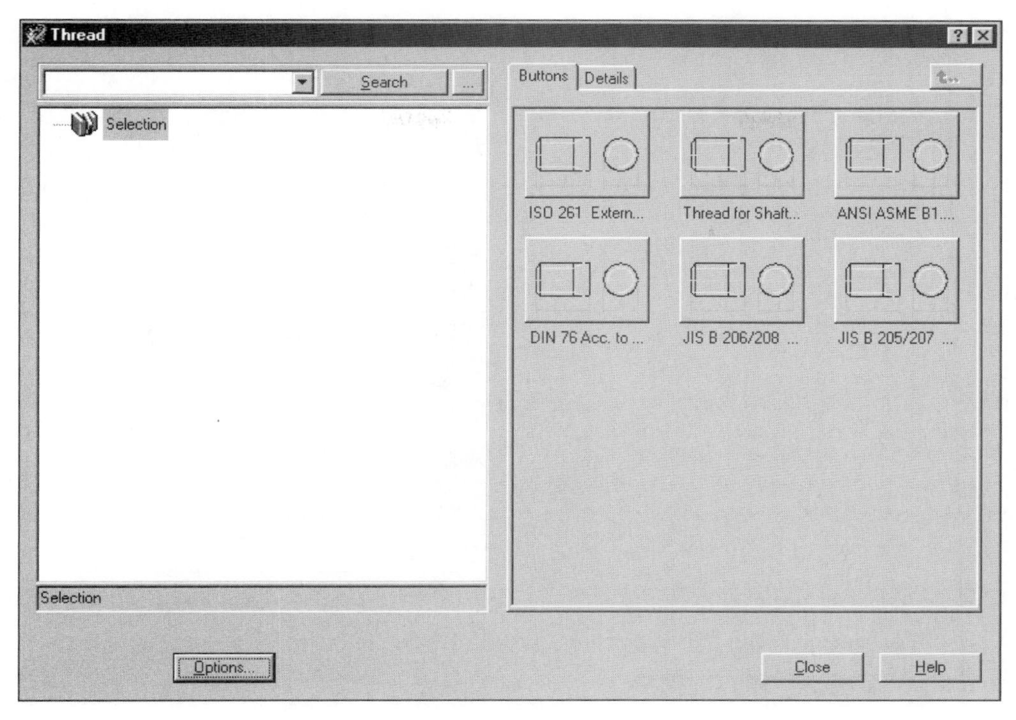

Figure 6–33

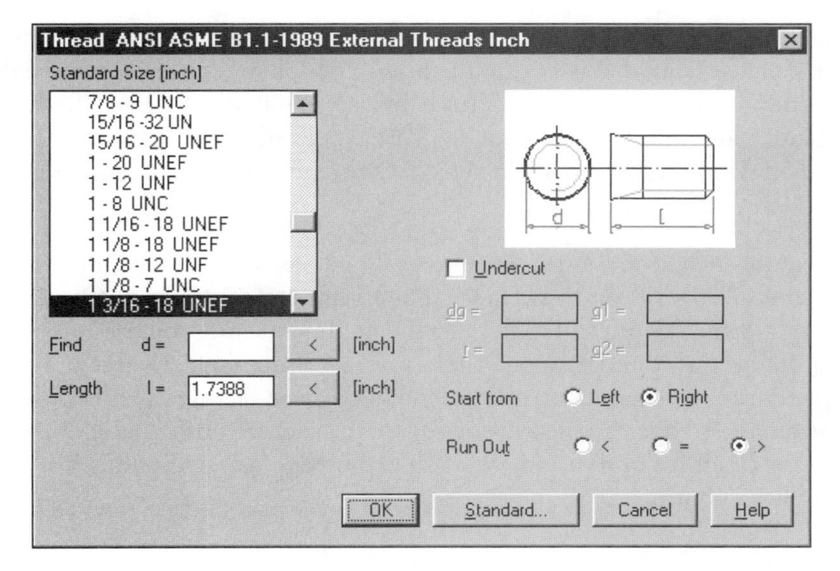

Figure 6–34

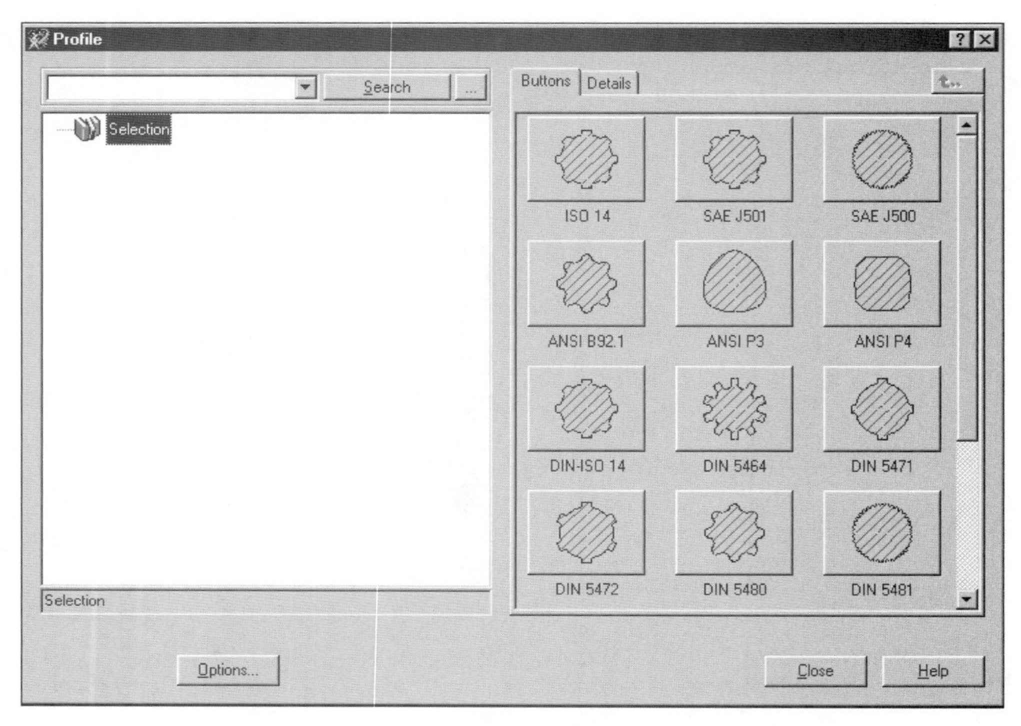

Figure 6–35

Once the desired option is selected, the Splined Shaft dialog box appears. Figure 6–36 shows the Splined Shaft dialog box for the SAE J501 spline. You can select the desired size from the list and accept it, as defined by the standard. Or select the Modified Design button and edit to meet your standards. Selecting the button brings up the Splined Shaft Modified Design dialog box, shown in Figure 6–37. Figure 6–38 shows the complete Profile.

Once a feature such as Cylinder, Profile, or Cone has been created, the button for Chamfer and Groove becomes available. The Chamfer option prompts you to select an edge to add the chamfer to. Then, at the command line, you are prompted for values to define the chamfer. Once these values are defined, the last prompt asks whether to create the chamfer as a chamfer or a revolved feature. This is important if there are multiple edges at the end of the shaft, such as with the above example Profile. The Chamfer option only works on the edge selected, while the Revolve option applies to both.

Splined Shaft SAE J501
Standard Size [inch]
3/4"
7/8"
1"
1-1/8"
1-1/4"
1-3/8"
1-1/2"
1-5/8"
1-3/4"
2"
2-1/4"
B
d
L
D
Find d =
<
[inch]
Length l =
2
<
[inch]
Modified Design
Standard...
OK
Cancel
Help

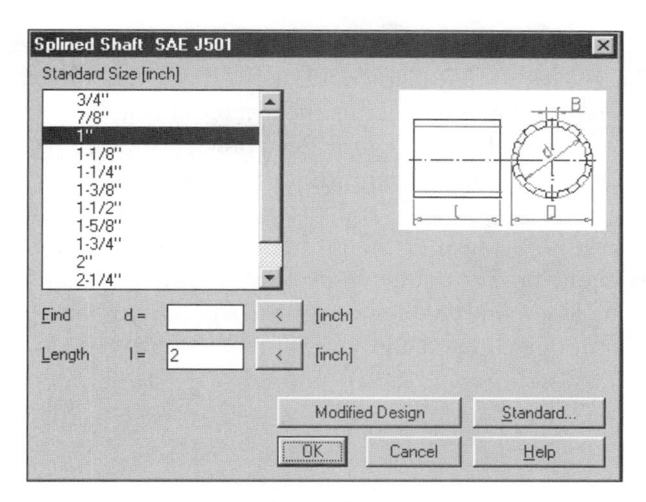

Figure 6–36

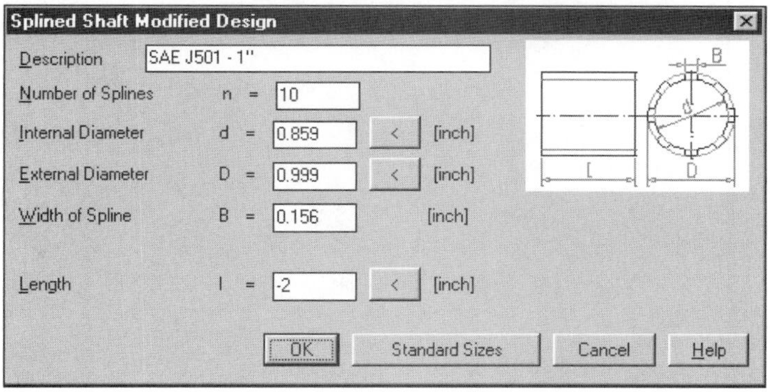

Splined Shaft Modified Design
Description
SAE J501 - 1"
Number of Splines n = 10
Internal Diameter d = 0.859 < [inch]
External Diameter D = 0.999 < [inch]
Width of Spline B = 0.156 [inch]
Length l = -2 < [inch]
B
d
L
D
OK
Standard Sizes
Cancel
Help

Figure 6–37

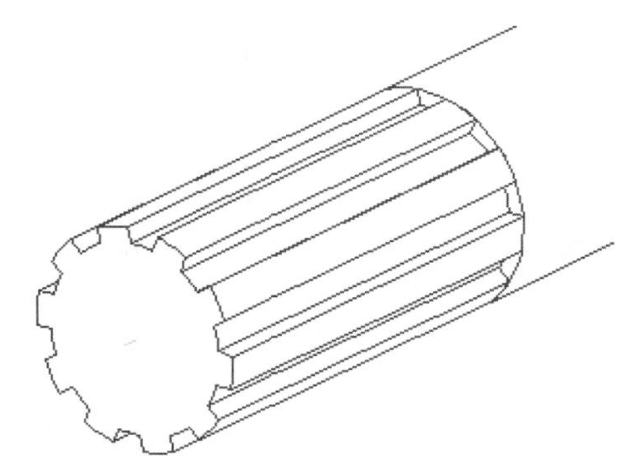

Figure 6–38

After pressing the Groove button you are prompted to select an existing cylinder or cone to apply the groove to. Mechanical Desktop then works from the end of the cylinder or cone and prompts you to define the distance from the end of the feature to place the groove. Mechanical Desktop provides a preview for the placement, as shown in Figure 6–39. Once the approximate location has been specified on screen, the direction of the groove is specified, and the exact distance can be entered at the command line. Enter the length of the groove next. Mechanical Desktop prompts you with: Specify diameter. The diameter specified is the inside diameter of the groove and must be a value less than the diameter of the shaft. Figure 6–40 shows a simple shaft with chamfer at one end and a groove.

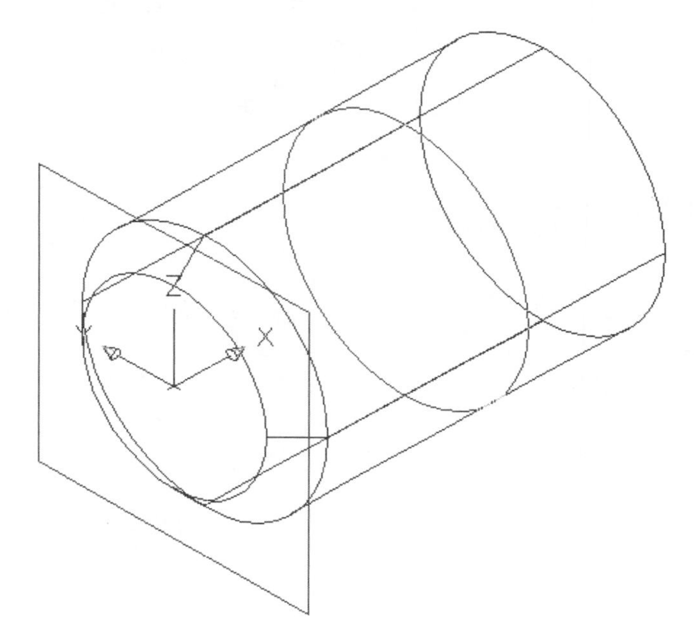

Figure 6–39

The first two buttons on the second row, Cylinder and Slope 1:x, may be more desirable to use when creating Cylinder and Cone sections of a shaft. Notice that the buttons have nearly the same picture as their counterparts above, with the addition of dimension values. When using these options, Mechanical Desktop prompts you for input values at the time the feature is created. This removes the need to use the Edit button after the feature has been created.

The Wrench button brings up the Wrench Opening dialog box, as shown in Figure 6–41. As with other features, the available options for Wrench Openings are determined by the standards installed.

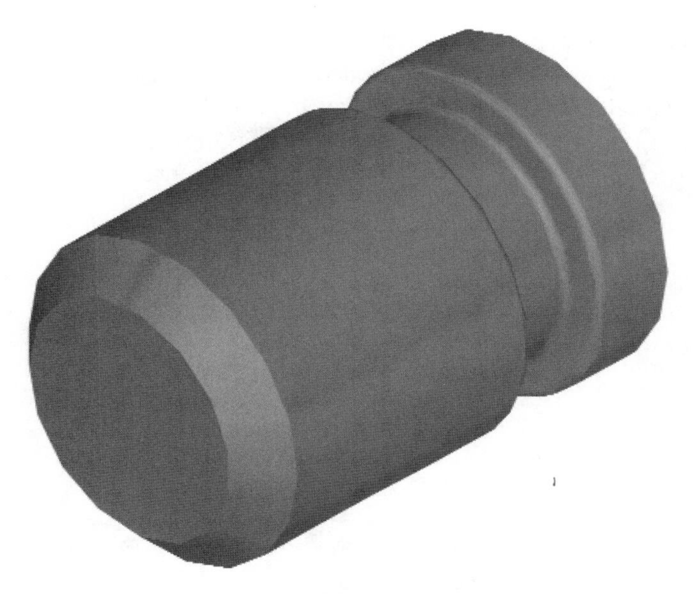

Figure 6–40

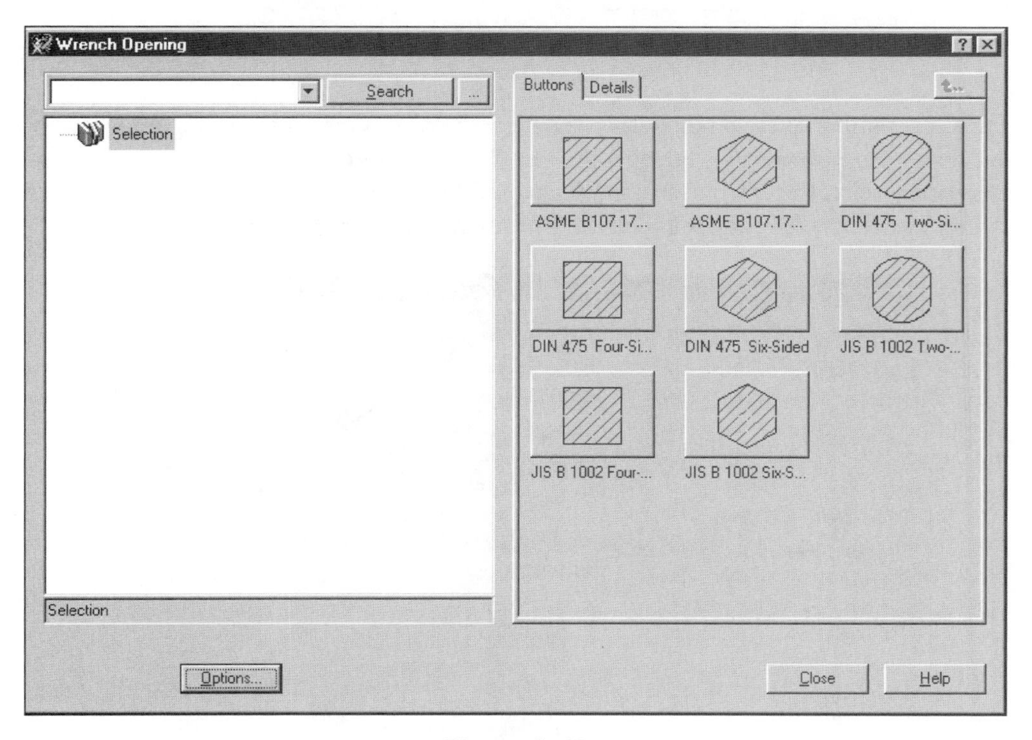

Figure 6–41

Once the specified wrench opening type has been selected, the Wrench Size dialog box appears, allowing you to select the wrench opening size (see Figure 6–42). Specify the length, and angle desired, and the Power Pack adds this feature and returns you to the 3D Shaft dialog box to add other features. The angle option rotates the wrench opening the specified number of degrees based on the centerline of the shaft.

Wrench Size ASME B107.17M Six-Sided
Standard Size [inch]
9/16"
5/8"
11/16"
3/4"
13/16"
7/8"
15/16"
1"
1 1/16"
1 1/8"
1 3/16"
Angle
0
[deg]
Find
s =
[inch]
Length
l =
2
[inch]
OK
Standard...
Cancel
Help

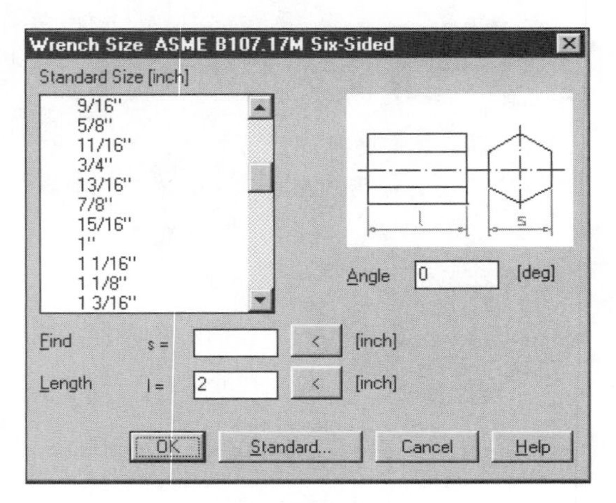

Figure 6–42

The Gear button takes you directly to the Gear dialog box, as shown in Figure 6–43. In this dialog box you specify all the parameters that will define the gear. Once those values are entered, select the OK button and Mechanical Desktop creates the gear. As mentioned earlier, by default, only one full tooth and two halves, appears (see Figure 6–44).

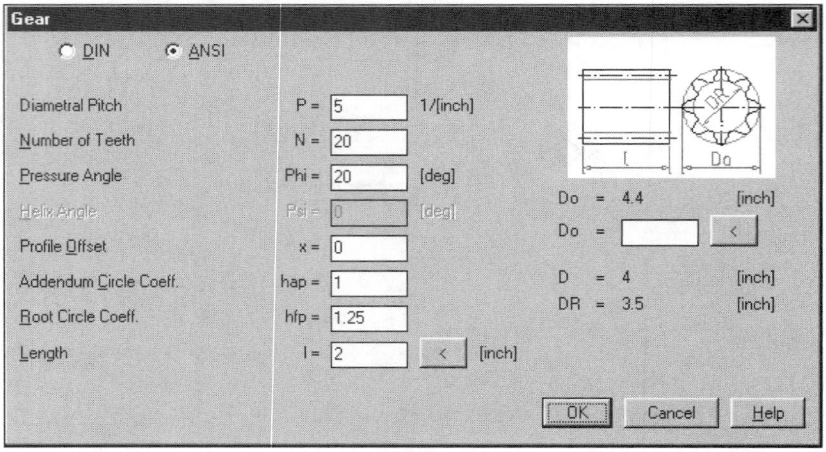

Gear
DIN
ANSI
Diametral Pitch
P = 5
1/[inch]
Number of Teeth
N = 20
Pressure Angle
Phi = 20
[deg]
Helix Angle
Psi = 0
[deg]
Profile Offset
x = 0
Addendum Circle Coeff.
hap = 1
Root Circle Coeff.
hfp = 1.25
Length
l = 2
[inch]
Do = 4.4 [inch]
Do =
D = 4 [inch]
DR = 3.5 [inch]
OK
Cancel
Help

Figure 6–43

Figure 6–44

If data is entered that does not match the information based on the standard, a message appears indicating the sketch for the gear cannot be completed (see Figure 6–45). Select OK to return to the Gear dialog box to adjust your values.

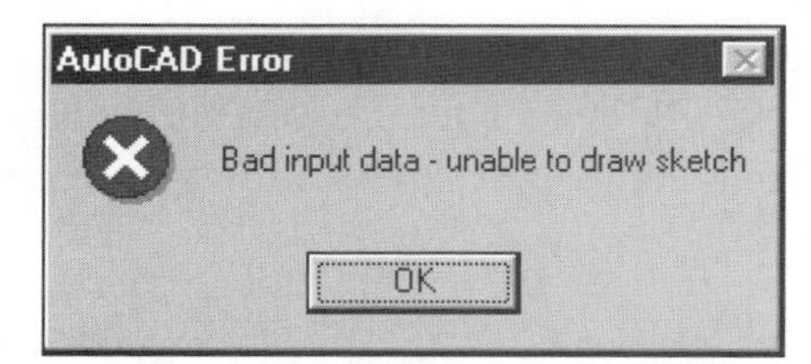

Figure 6–45

The Fillet button works much the same as the AMFILLET command and allows you to radius an edge of the shaft. The major difference is that the Fillet button from the Gear dialog box analyzes the selected edge and reports whether it can have a radius applied. If this is possible, it reports the maximum value that can be entered (see Figure 6–46).

```
Not possible to create fillet
Select edge for radius:
Specify radius (max. 0.202) or  [Associate to/Equation assistant] <.1250>:
Target: SHAFT GENERATOR  -1.4933, -2.7392, 0.0000    SNAP GRID ORTHO POLAR OSNAP OTRACK LWT MODEL
```

Figure 6–46

The Right Inner Contour and Left Inner Contour tabs are identical in operation. The difference between them is which end of the shaft will be affected by the commands. Figure 6–47 shows the Left Inner Contour tab. The function of the buttons on both of the Inner Contour tabs is identical to their counterparts on the Outer Contour tab.

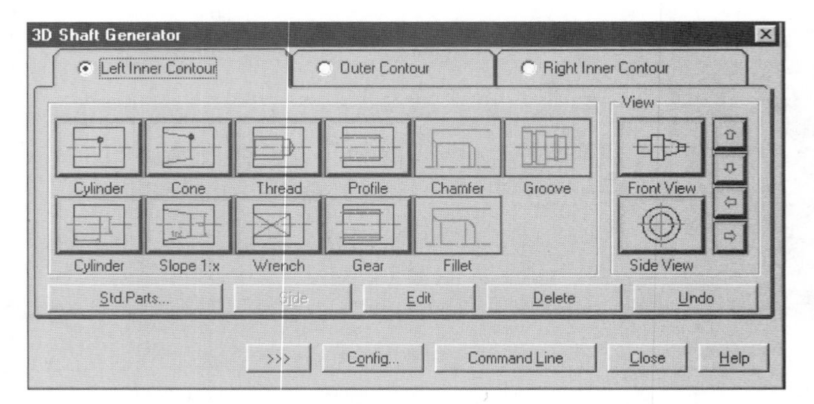

Figure 6–47

STEEL SHAPES

The Mechanical Desktop Power Pack also includes predefined Steel Shapes such as I-beams, angle brackets, box steel in both solid and hollow styles, and others (see Figure 6–48). Just as in other Power Pack Features, the Steel Shapes available are based on the Standards installed. Figure 6–49 shows some of the different types of I-beam shapes available.

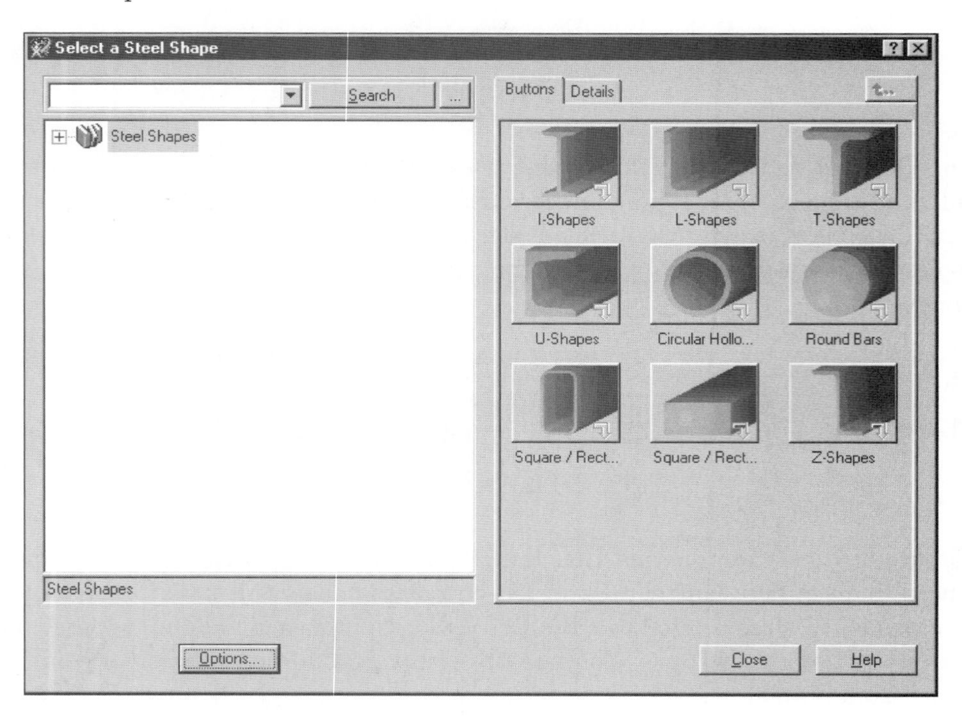

Figure 6–48

Once a specific type of I-beam has been selected, Mechanical Desktop prompts you to define the location for the shape based on two selected points: concentric to a cylinder, or relative to a cylinder. Once the location is defined, a dialog box specific to that type appears, allowing you to select the standard size. Figure 6–50 shows the ANSI HP - Size Selection dialog box.

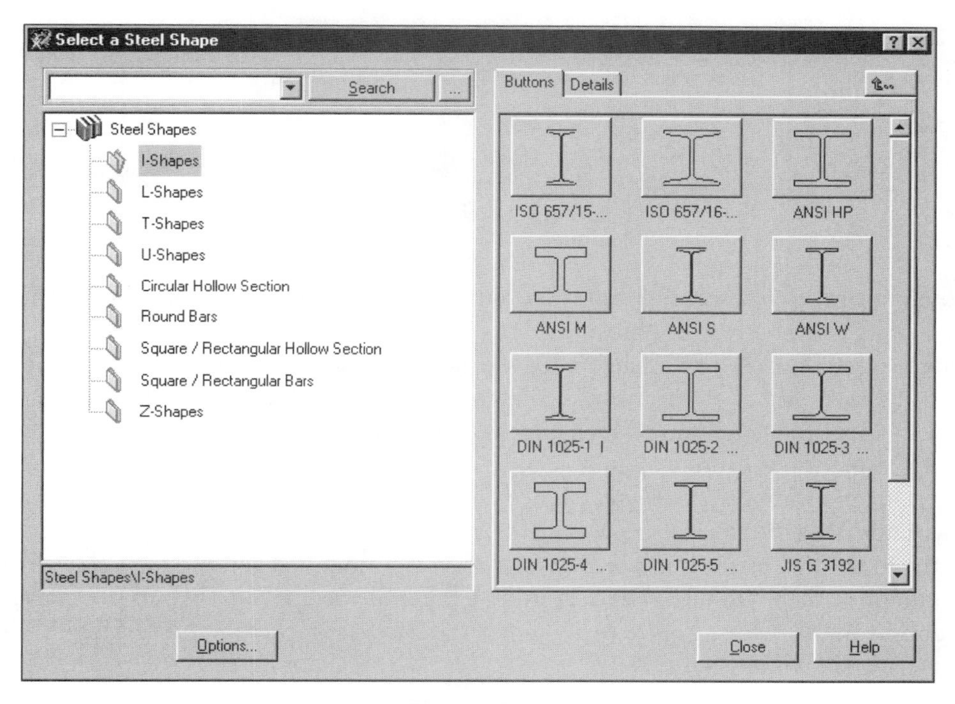

Figure 6–49

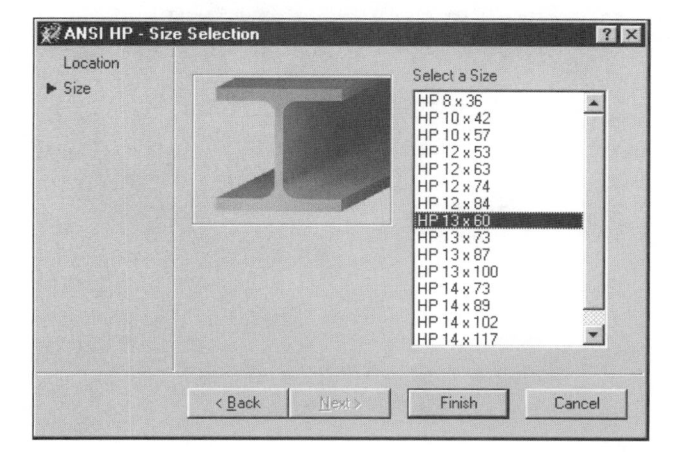

Figure 6–50

Once the size has been selected, Mechanical Desktop returns to the graphics screen and allows you to drag the cursor to define the length. You can base the length on an existing part, or type in a specific value, or select the dialog option. Steel Shapes also give you the option to Associate to another feature. Figure 6–51 shows the command line sequence for using the Associate to option.

```
Command: _amstlshap3d
Select first point [Concentric/cYlinder/two Edges]:
Select second point [Concentric/cYlinder/two Edges]:
Drag size [Dialog/Associate to/Equation assistant]: a
Select feature or dimension to associate to:
Select feature or dimension to associate to or [Undo]:
Command:
Target: SHAFT GENERATOR -23.2687, 41.0960, 0.0000    SNAP GRID ORTHO POLAR OSNAP OTRACK LWT MODEL
```

Figure 6–51

When prompted to Drag size, enter the letter A at the command line, or right-click in the graphics area and select Associate to from the menu. Mechanical Desktop prompts you to select a feature or dimension to associate to. You can select a dimension if one is available, or select a feature. The dimensions used to define the selected feature appear (like edit sketch) and Mechanical Desktop again prompts you to select a feature or dimension. Once the dimension is selected, the Steel Shape is created at the length of the dimension selected. Once complete, (see Figure 6–52) other features can be added to the steel shape. While the Steel Shape can be edited by right-clicking on the individual features in the Browser, it is recommended you use the Power Edit command, which will be covered later. The Steel Shape appears with its own designation in the Browser. Figure 6–53 shows the Browser representation of the ANSI HP shape shown in Figure 6–52.

Figure 6–52

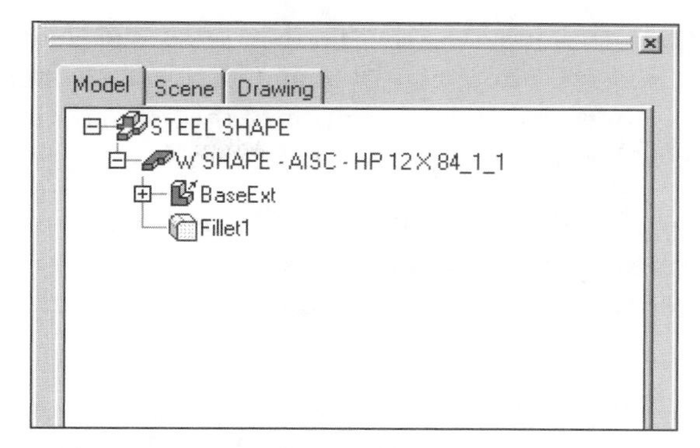

Figure 6–53

POWER COMMANDS

Mechanical Desktop Power Pack comes with its own Power commands. The Power commands are located on the Mechanical Main toolbar (see Figure 6–54). The Power commands include: Power Dimensioning, Power Edit, Power Copy, Power Erase, and Power Snap, as shown in Figure 6–55.

Figure 6–54

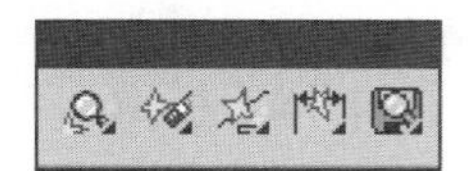

Figure 6–55

The Power commands incorporate more intelligence than their standard Mechanical Desktop counterparts, especially when dealing with entities created with Power Pack commands. For example, look at the Steel Shape that was just created. You could use AMEDITFEAT to change the extrusion depth, or edit the feature. However, by using the Power Edit (AMPOWEREDIT) command, you are taken back to the dialog box where the Steel Shape was defined, allowing you to change to a different size, and redefine the length. When Power Edit is used on Mechanical Desktop features, it just reverts back to the appropriate dialog box that created the feature, just as AMEDITFEAT would do.

The Power Erase command (AMPOWERASE) differs from the standard ERASE command when used on parts and features created with the Power Pack. For example, in the case of a Screw Connection, using the standard ERASE command requires that all components of the Screw Connection be selected to delete them. Also, you are prompted whether or not to delete the definition of the parts. Using Power Erase will delete the entire screw connection by selecting any single component.

The Power Copy command (AMPOWERCOPY) works much like Power Erase. The selection of any part of a Screw Connection selects the entire Screw Connection. When Power Copy or Power Erase is used on Mechanical Desktop parts; it just reverts back to the standard COPY or ERASE command.

Power Dimensioning (AMPOWERDIM) and Power Snap (AMPOWERSNAP) commands are more involved than the other Power commands. Power Dimensioning is used in all aspects of dimensioning in Mechanical Desktop. Both Parametric and Reference dimensions are placed with the Power Dimensioning command. Power Snap allows you to have multiple snap settings available.

Power Dimensioning allows you to create many different types of dimensions. When invoked, the Power Dimensioning command contains options for Angular, Baseline, and Chain dimensions. The Power Dimension command can also access other options and can update existing dimensions. Whether you are Power Dimensioning on sketches or in drawing views, the procedure is the same. The command line prompts you to: Specify first extension line origin. Or, as shown in Figure 6–56, select one of the command line options.

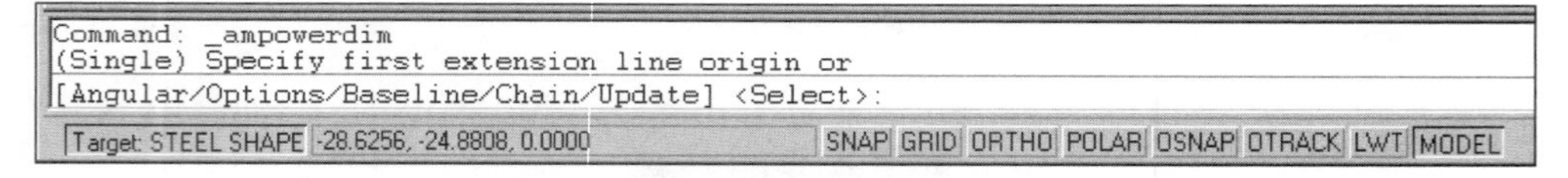

Figure 6–56

Once the dimension origins and location are specified, the Power Dimensioning dialog box appears (see Figure 6–57). The Power Dimensioning dialog box has three tabs: General, Geometry, and Units. It also has two buttons at the top of the dialog box: Add Fit and Add Tolerance. There are three buttons at the bottom: OK, Cancel, and Help.

The General tab contains three fields to input information, and seven buttons. At the top of the General tab, is a field with the expression <<xx>> contained inside. That expression is the parametric value and should *not* be edited. However, in this field, any prefix or suffix can be added and displayed on the dimension. The input field below this is called Expression. Here, specific values are entered for the parametric dimensions. When placing reference dimensions in drawing views, this field is labeled Exact Distance, and while it shows the value, the field itself will be grayed out. To the right of the Expression field is the Precision field. This field specifies the number of decimal places to be displayed by the dimension. Above these fields are five buttons that control different options for dimensioning. From left to right they are: Dimension Symbol, Alternate Unit Symbol, Underline Dimension Text, Box Dimension Text, and Special Characters.

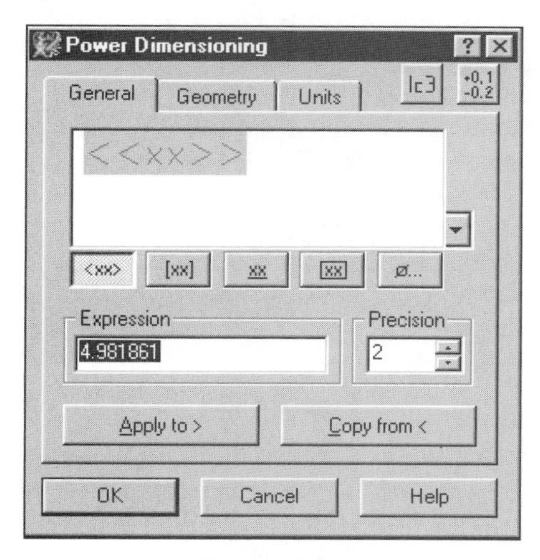

Figure 6–57

The Apply to > button at the bottom of the General tab, is used to update dimensions. Selecting this button brings up the Select Parameters to Copy dialog box shown in Figure 6–58. You can use this to apply any, or all categories to selected dimensions. The Copy from < button copies the same parameters from an existing dimension to the dimension about to be placed.

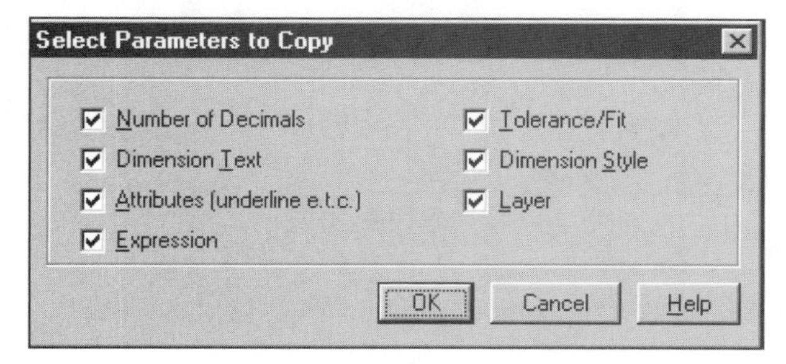

Figure 6–58

The Geometry tab allows us to edit the offset distance of the text from the dimension line (see Figure 6–59).

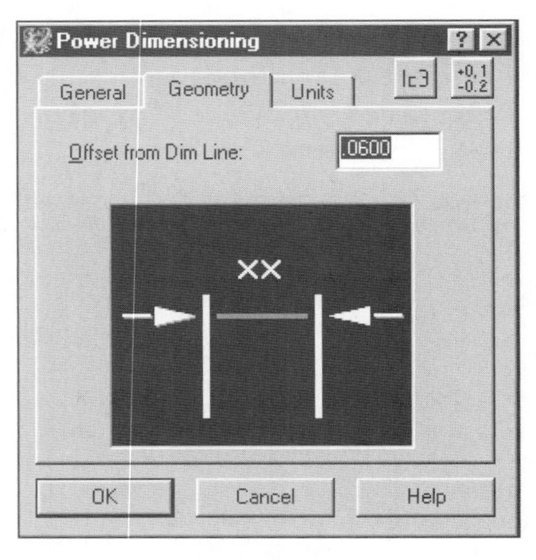

Figure 6–59

The Units tab consists of two sections. The Primary section allows you to choose the type of Unit to be used, the Linear Scaling, and the Round Off value if desired. The Alternate section allows you to set the number of decimal places for alternate units, if they have been used (see Figure 6–60).

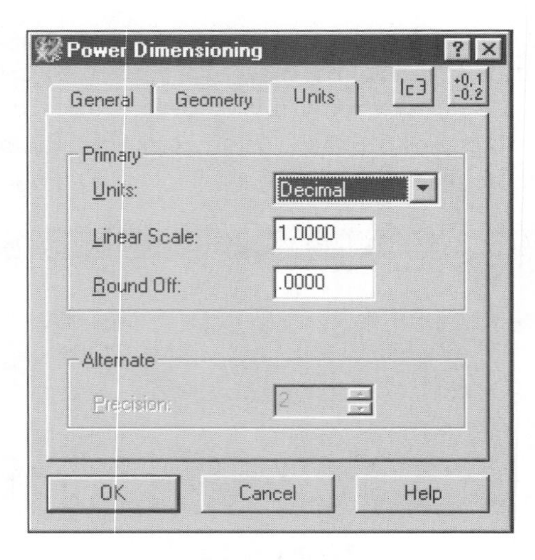

Figure 6–60

The Add Fit and Add Tolerance buttons expand the Power Dimensioning dialog box to include options for these two types of dimensions. The Add Fit button, shown in Figure 6–61, allows you to specify symbol options and to control Primary and Alternate unit precision. The Add Tolerance button, shown in Figure 6–62, allows you to set Upper and Lower Deviation limits, and Primary and Alternate unit precision.

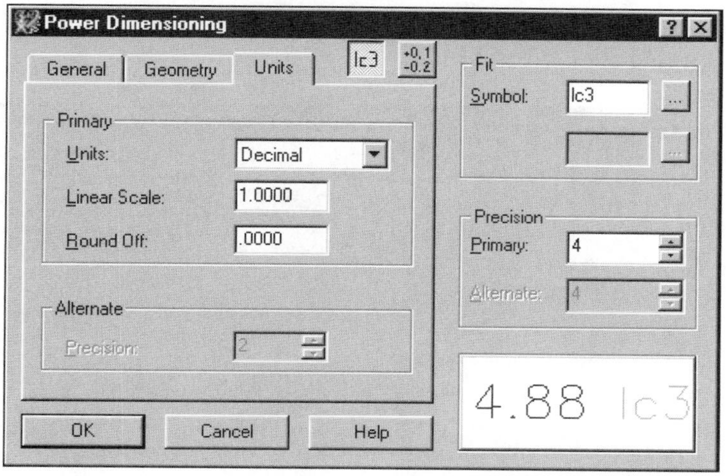

Power Dimensioning
General
Geometry
Units
lc3
+0.1
-0.2
Primary
Units: Decimal
Linear Scale: 1.0000
Round Off: .0000
Alternate
Precision: 2
Fit
Symbol: lc3
Precision
Primary: 4
Alternate: 4
4.88 lc3
OK
Cancel
Help

Figure 6–61

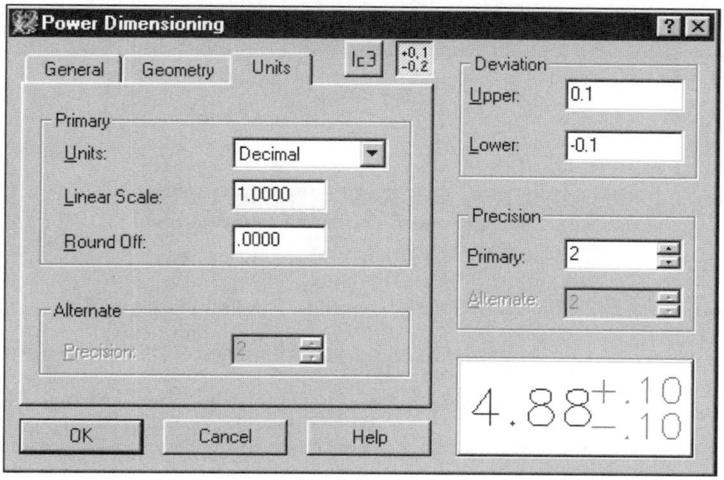

Power Dimensioning
General
Geometry
Units
lc3
+0.1
-0.2
Primary
Units: Decimal
Linear Scale: 1.0000
Round Off: .0000
Alternate
Precision: 2
Deviation
Upper: 0.1
Lower: -0.1
Precision
Primary: 2
Alternate: 2
$4.88^{+.10}_{-.10}$
OK
Cancel
Help

Figure 6–62

Power Snap allows you to configure the four different Object Snap tabs found in the Power Snap Settings dialog box (see Figure 6–63). The settings are the same as the normal object snap settings. However, you can have four different configurations of object snaps running in a single drawing file. To set the different object snap modes, click on a settings tab and set the desired object snap. You repeat the process up to four times for the different configurations needed. To change to a different setting, select the Power Snap command and select the tab that contains the desired object snaps. This is especially useful now that object snap settings are no longer held in the drawing file.

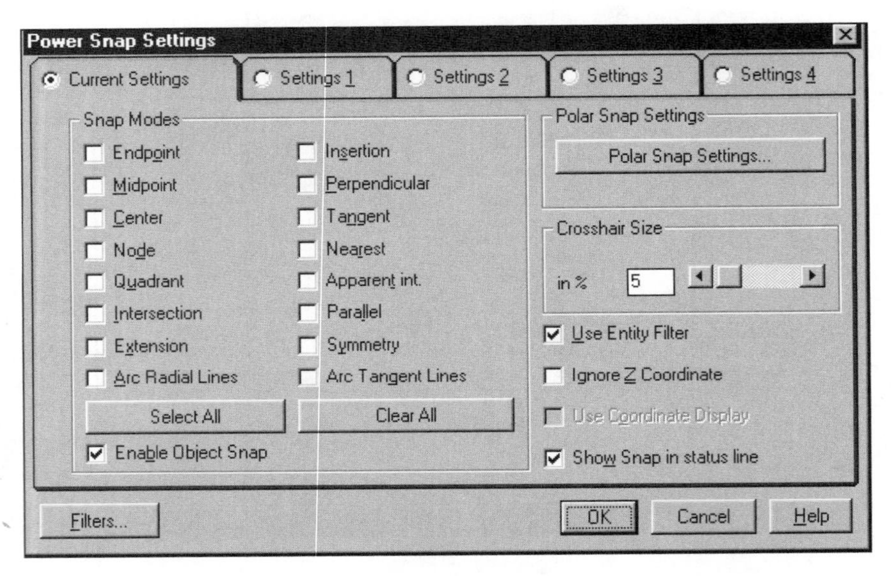

Figure 6–63

Tutorial 6–1 Creating Holes with the Power Pack

1. If not already running, start Mechanical Desktop 5.
2. Open file MD06-EX 1.dwg. The file should look like Figure 6–4.

   Using the Power Pack, you will place four ¼" holes on all four corners.
3. From the Content 3D pull-down menu, select Holes, then Through Holes.

   If not already loaded, the Power Pack portion of Mechanical Desktop will load at this point.
4. From the Select a Through Hole dialog box, select the Free option.
5. From the Hole Position Method First Hole dialog box, select the 2 Edges method.

Figure 6–64

6. Select the two edges to the far left. Place the hole 0.5 inches from each edge.
7. Accept the default <Thru> termination option.
8. From the Free – Nominal Diameter dialog box, select the ¼" option as shown in Figure 6–65. Select the Finish button.

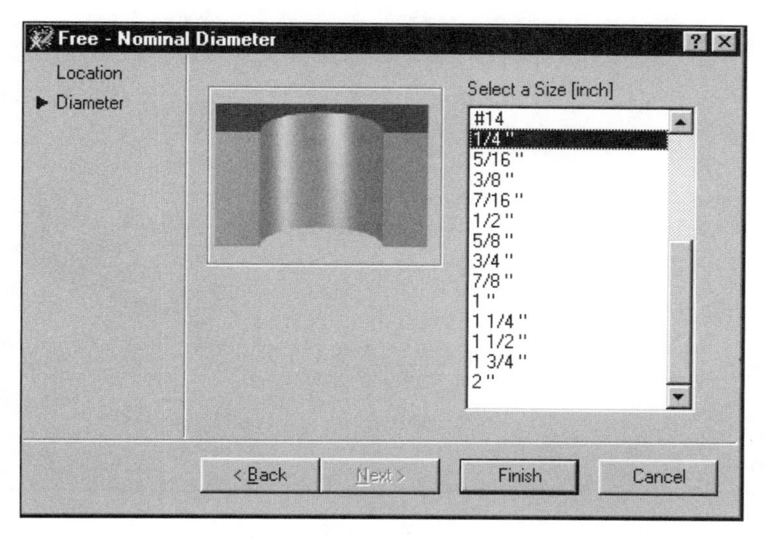

Figure 6–65

9. In the Browser, right-click on the newly created AM_HOLE and select Edit. Notice the Drill Size diameter of .266.
10. Repeat the process for the other four corners.

Tutorial 6–2 Fasteners

1. If not already running, start Mechanical Desktop 5.
2. Start a drawing from scratch.
3. From the Content 3D menu select Fasteners then Screws.
4. In the Select a Screw dialog box, select the Socket Head Types button.
5. Select the Socket Head Cap Screw UNC as shown in Figure 6–66.

Note: The tooltip on your screen may differ from the figure depending on the Standards loaded.

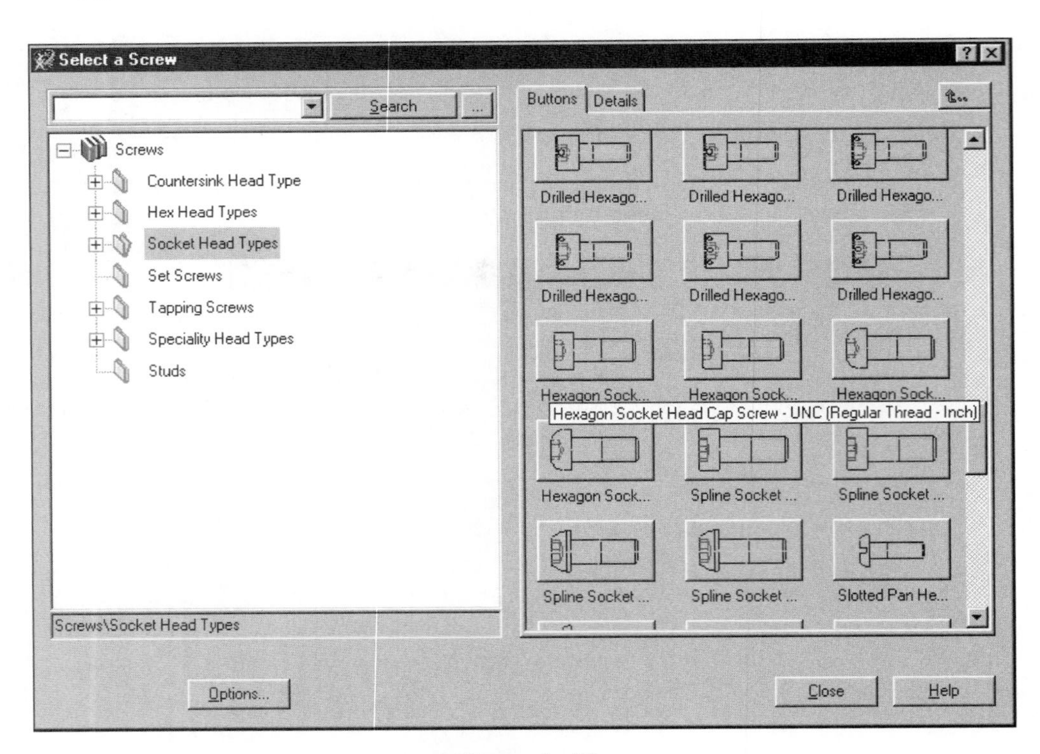

Figure 6–66

6. When prompted, select two points in the graphics area, with the second point directly below the first point.
7. Select the ½"-13 UNC option from the Hexagon Socket Head Cap Screw dialog box. Press the Finish button.
8. When the command line prompts: Drag size [Dialog], right-click in the graphics area and select Dialog from the menu.
9. In the Select a Row dialog box, select the ½-13 x 1-1/2 row, as shown in Figure 6–67.

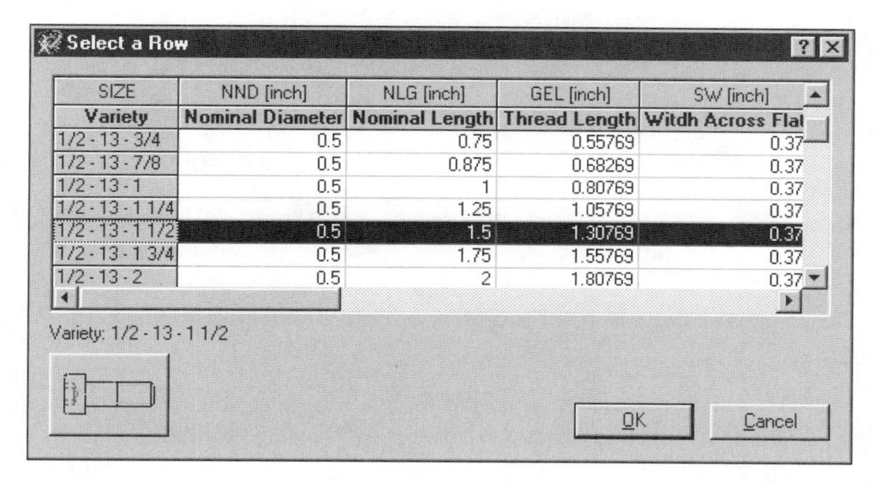

SIZE	NND [inch]	NLG [inch]	GEL [inch]	SW [inch]
Variety	Nominal Diameter	Nominal Length	Thread Length	Witdh Across Flat
1/2 - 13 - 3/4	0.5	0.75	0.55769	0.37
1/2 - 13 - 7/8	0.5	0.875	0.68269	0.37
1/2 - 13 - 1	0.5	1	0.80769	0.37
1/2 - 13 - 1 1/4	0.5	1.25	1.05769	0.37
1/2 - 13 - 1 1/2	0.5	1.5	1.30769	0.37
1/2 - 13 - 1 3/4	0.5	1.75	1.55769	0.37
1/2 - 13 - 2	0.5	2	1.80769	0.37

Figure 6–67

10. Press the number 8 key and press ENTER to go to an isometric view. Your screen should resemble Figure 6–68.

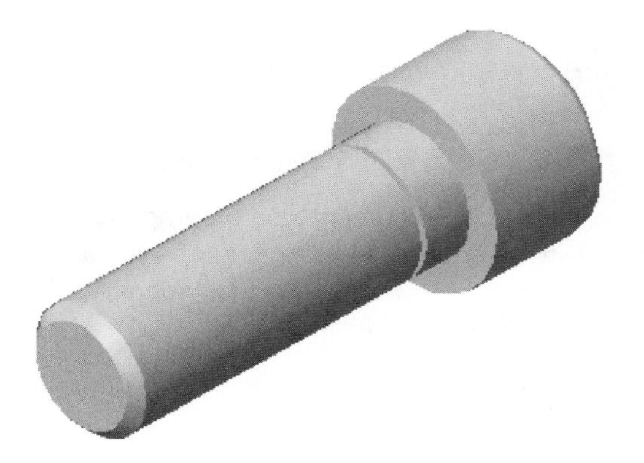

Figure 6–68

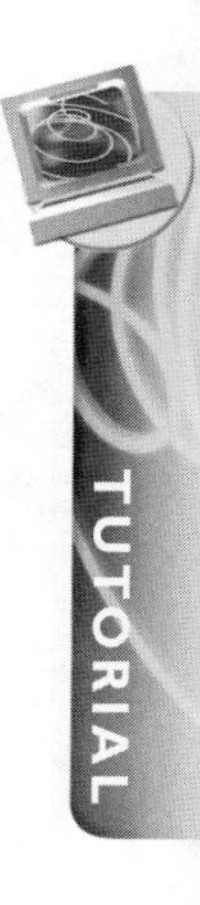

Tutorial 6–3 Screw Connections

1. If not already running, start Mechanical Desktop 5.
2. Open file MD06-EX 2.dwg. The file should look like Figure 6–9.

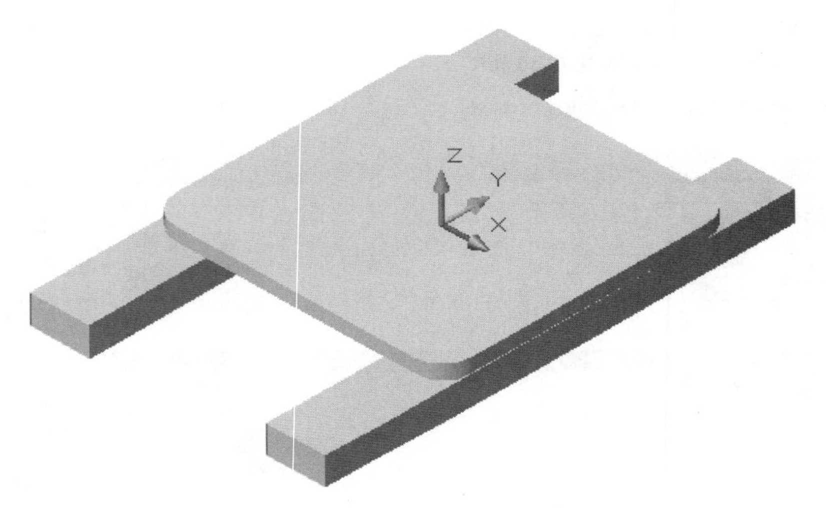

Figure 6–69

3. The Plate is ¼" thick, and the two straps below are ½" thick. From the Content 3D pull-down menu, select Screw Connection. You will place a 3/8"-16 Socket Head Cap.
4. For the Socket Head Cap Screw, make the selections in the Screw Connections – 3D dialog box, as shown in Figure 6–70. When the settings are complete, press the Finish button.
5. Use the concentric option to place the first hole. Select the top edge of the fillet to be concentric.
6. Right-click and select Thru as the termination option.
7. When this is completed, the Hole Position Method dialog box will appear for placement of the second hole. Use the 2 Edges option and select the top two edges on the rectangular part. Mechanical Desktop will calculate the location and display the result on the command line.
8. Again, select the Thru option.
9. In the Select a Row dialog box, select OK.
10. The Screw Connection will calculate the length needed and round to the nearest standard length to put the Screw Connection through both parts and add the washer and nut. Your file should resemble Figure 6–71.

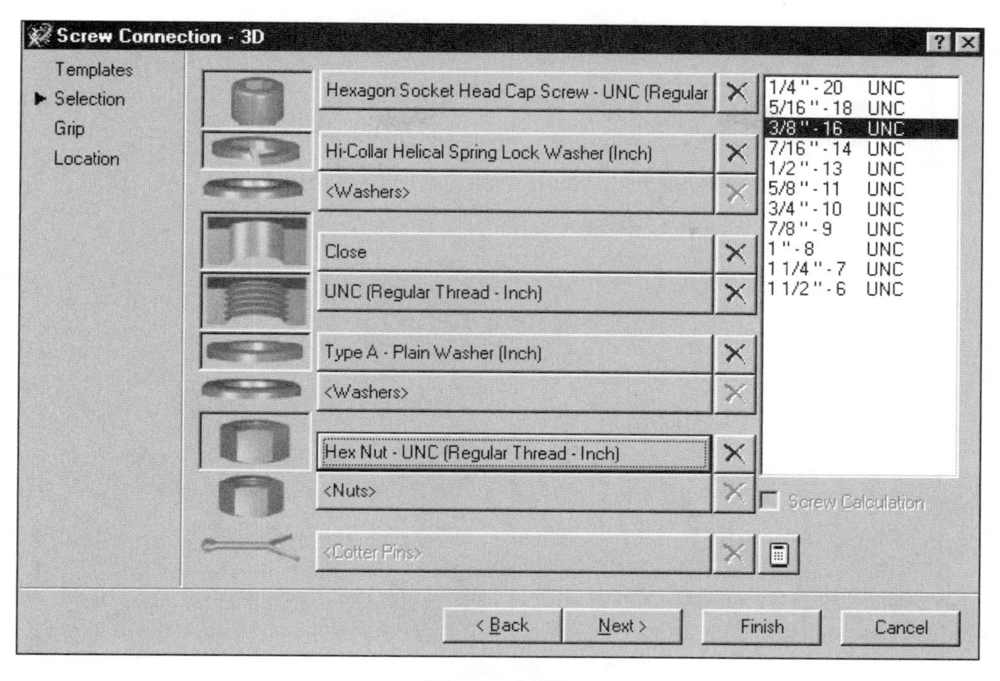

Figure 6–70

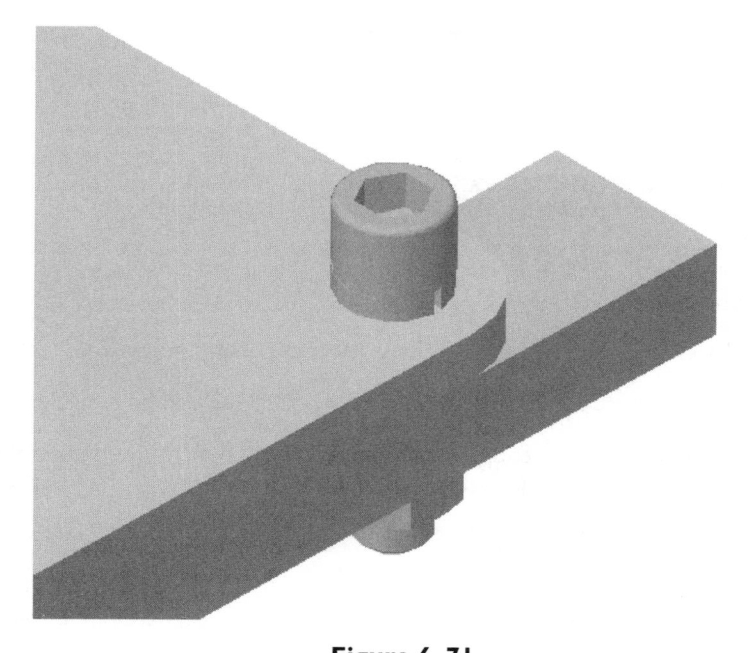

Figure 6–71

11. Place three more Screw Connections to complete the part.

Note: You may want to use the Screw Connection just to place the hole, and create a sub assembly of the existing Screw Connection to use assembly constraints.

12. Figure 6–72 shows completed assembly. Save the file.

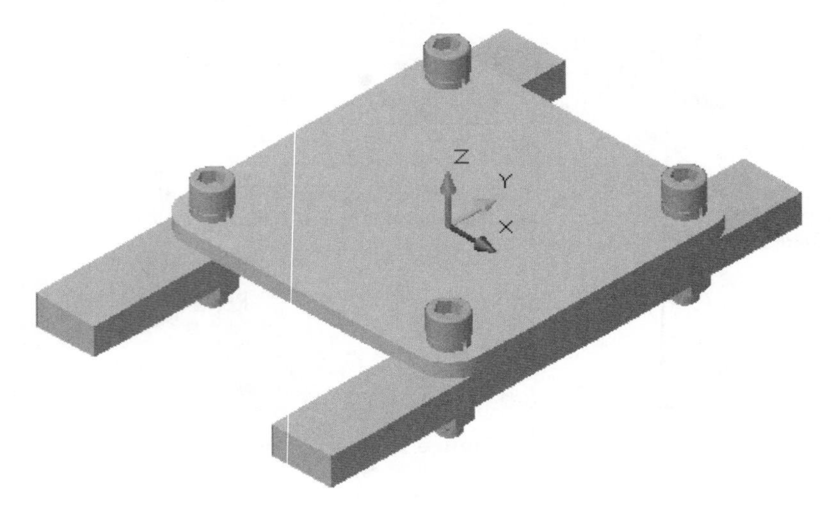

Figure 6–72

Tutorial 6–4 Shaft Generator

1. If not already running, start Mechanical Desktop 5.
2. Start a new file.
3. From the Content 3D pull-down menu, select the Shaft Generator.
4. At the Specify start point prompt, select any point in the graphics area.
5. At the Specify centerline endpoint prompt, select a point to the right of the previous point. You may have to turn Ortho or Polar on.
6. When prompted to Specify point for new plane <parallel to UCS>, select any point above the centerline previously defined.
7. When the 3D Shaft Generator dialog box appears, select the Cylinder button from the second row.

TUTORIAL

8. On the command line, specify a length of 1.5, and a diameter of 1.
9. When the dialog reappears, select the Slope 1:x button from the second row.
10. Specify a length of 1.5, a starting diameter of 1, and an ending diameter of 1.25.
11. When the dialog reappears, select the Gear button in the second row.
12. Enter a length of .5 and accept the rest of the defaults from the Gear dialog box.
13. When the dialog reappears, select the Profile button in the top row.
14. From the Profile dialog box select the SAE J501 button.
15. In the Splined Shaft dialog box, select the 1-1/4" size and press OK.
16. In the 3D Shaft Generator dialog box, select the Wrench button on the top row.
17. Select the AMSE B107 button or other applicable Wrench option.
18. In the Wrench Size dialog box, select the 1" size and press OK.
19. In the 3D Shaft Generator dialog, select the Thread button in the top row.
20. In the Thread dialog, select the ANSI ASME button.
21. In the next dialog box, select the 1-8 UNC option and press OK.
22. In the 3D Shaft Generator dialog, select the Chamfer button in the top row.
23. When prompted, select the first edge.
24. Enter a length of .25, and an angle of 45.
25. Select the Chamfer option.
26. Use the buttons under the View section to rotate the shaft into an ISO view with the first cylinder created, closest to you.
27. Select the Fillet button in the second row.
28. Select the edge between the Slope and the Gear.
29. On the command line, enter a value of .25, and then choose the Fillet option.
30. In the 3D Shaft Generator dialog, select the Left Inner Contour tab.
31. Select the Cylinder button on the second row.
32. Enter a value of 1 for the length, and .5 for the diameter.
33. In the 3D Shaft Generator dialog, press the Close button.
34. Your Shaft should resemble 6–73.

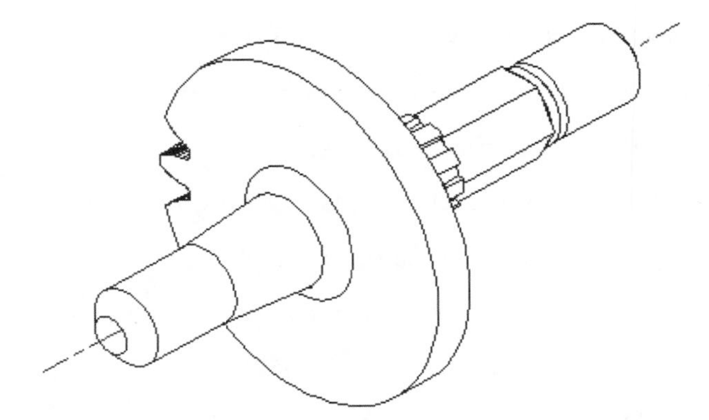

Figure 6–73

The previous exercise was not intended to create a working shaft, but to use as many of the options in the 3D Shaft Generator dialog box as possible.

Tutorial 6–5 Steel Shapes

1. If not already running, start Mechanical Desktop 5.
2. Start a new file.
3. From the Content 3D pull-down menu, select Steel Shapes.
4. In the Steel Shapes dialog box, select I-Shapes.
5. Select the ANSI HP option, as shown in Figure 6–74.
6. Pick two points, vertical to each other. You may want to turn on Polar or Ortho.
7. In the ANSI HP – Size Selection dialog box, select the HP 12 x 74 size, as shown in Figure 6–75. Press the Finish button.
8. When prompted to drag the size, right-click in the graphics area and select Dialog from the context menu.
9. In the Value column of the Enter Values dialog box, enter 30, as shown in Figure 6–76. Press OK.

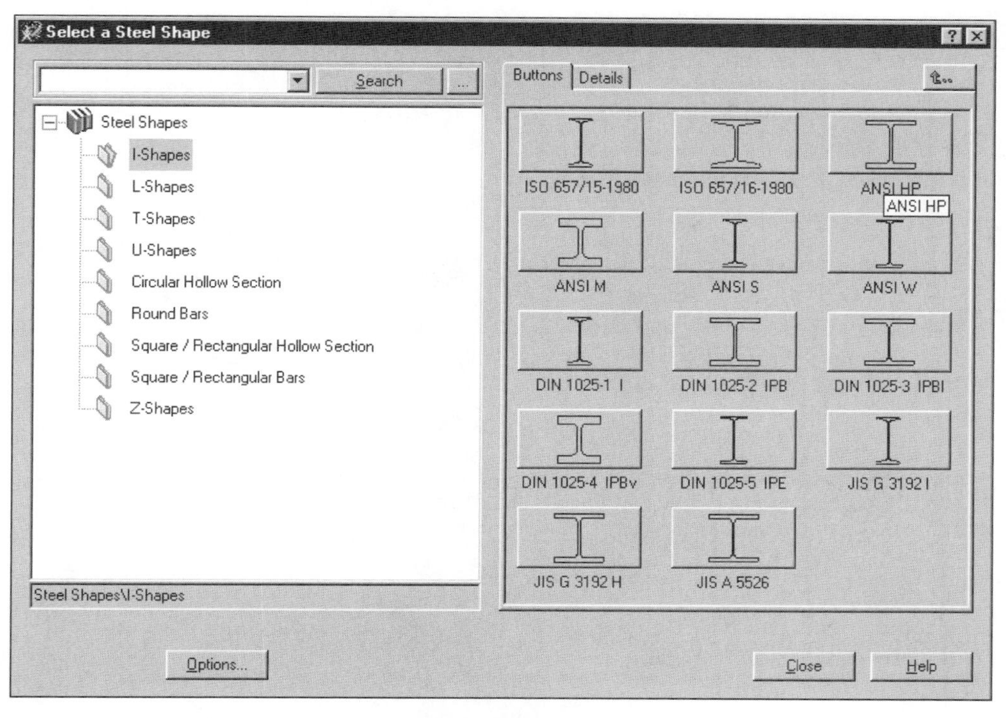

Figure 6–74

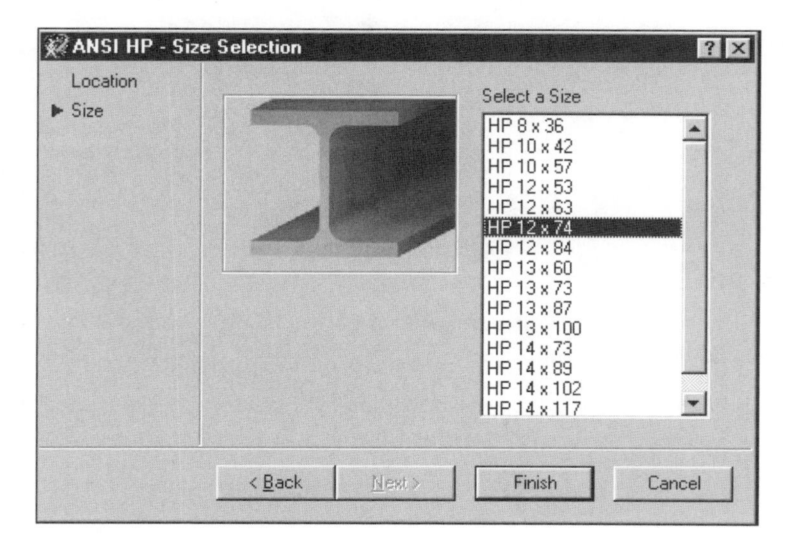

Figure 6–75

TUTORIAL

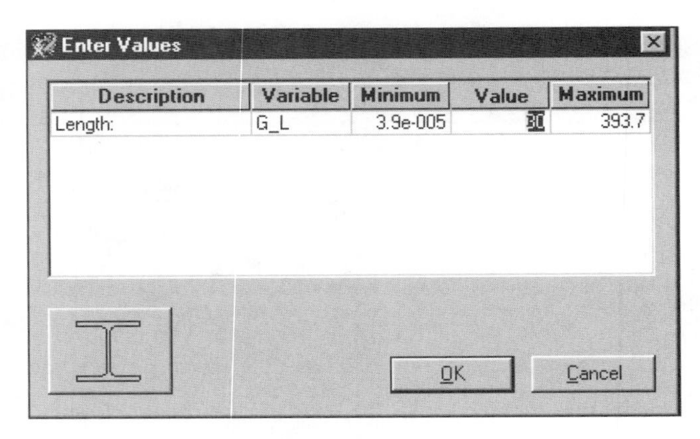

Figure 6–76

10. Press the number 8 key and press ENTER to go to an isometric view. Your file should resemble Figure 6–77.

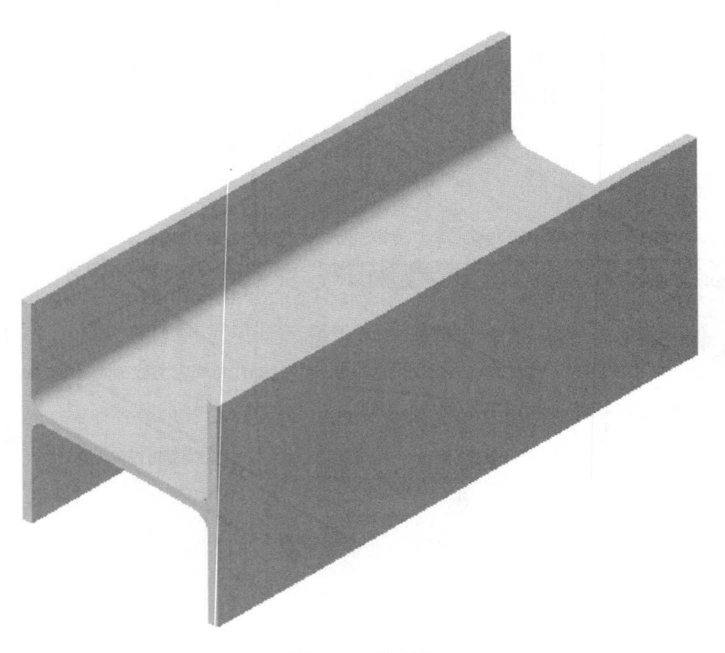

Figure 6–77

REVIEW QUESTIONS

1. Power Pack holes and Mechanical Desktop holes are the same. T or F?
2. The sizes available for selection are based on the current ________________.
3. Screw Connections are automatically constrained when located. T or F?
4. Power Pack holes have more placement options than standard Mechanical Desktop holes. T or F?
5. Fasteners are just a parts library that inserts the selected items. T or F?
6. Screw Connections allow you to place several components in a single command. T or F?
7. The inner contours in the Shaft Generator must be created before the outer contour is defined. T or F?
8. Standard Parts can be added from within the Shaft Generator dialog box. T or F?
9. Once a shaft is completed, it must be deleted to change any of the features. T or F?
10. AMPOWERERASE will only erase the single item selected of a Screw Connection. T or F?

INDEX

Note: Topics in SMALL CAPS indicate command-line or AutoCAD commands. Page numbers in **boldface** indicate material in figures.

B

C

D

E

F

G

H

I

L

R

S

T

U

LICENSE AGREEMENT FOR AUTODESK PRESS

THOMSON LEARNING™

Educational Software/Data

You the customer, and Autodesk Press incur certain benefits, rights, and obligations to each other when you open this package and use the software/data it contains. BE SURE YOU READ THE LICENSE AGREEMENT CAREFULLY, SINCE BY USING THE SOFTWARE/DATA YOU INDICATE YOU HAVE READ, UNDERSTOOD, AND ACCEPTED THE TERMS OF THIS AGREEMENT.

Your rights:

1. You enjoy a non-exclusive license to use the enclosed software/data on a single microcomputer that is not part of a network or multi-machine system in consideration for payment of the required license fee, (which may be included in the purchase price of an accompanying print component), or receipt of this software/data, and your acceptance of the terms and conditions of this agreement.
2. You own the media on which the software/data is recorded, but you acknowledge that you do not own the software/data recorded on them. You also acknowledge that the software/data is furnished "as is," and contains copyrighted and/or proprietary and confidential information of Autodesk Press or its licensors.
3. If you do not accept the terms of this license agreement you may return the media within 30 days. However, you may not use the software during this period.

There are limitations on your rights:

1. You may not copy or print the software/data for any reason whatsoever, except to install it on a hard drive on a single microcomputer and to make one archival copy, unless copying or printing is expressly permitted in writing or statements recorded on the diskette(s).
2. You may not revise, translate, convert, disassemble or otherwise reverse engineer the software/data except that you may add to or rearrange any data recorded on the media as part of the normal use of the software/data.
3. You may not sell, license, lease, rent, loan, or otherwise distribute or network the software/data except that you may give the software/data to a student or and instructor for use at school or, temporarily at home.

Should you fail to abide by the Copyright Law of the United States as it applies to this software/data your license to use it will become invalid. You agree to erase or otherwise destroy the software/data immediately after receiving note of Autodesk Press' termination of this agreement for violation of its provisions.

Autodesk Press gives you a LIMITED WARRANTY covering the enclosed software/data. The LIMITED WARRANTY can be found in this product and/or the instructor's manual that accompanies it.

This license is the entire agreement between you and Autodesk Press interpreted and enforced under New York law.

Limited Warranty

Autodesk Press warrants to the original licensee/ purchaser of this copy of microcomputer software/ data and the media on which it is recorded that the media will be free from defects in material and workmanship for ninety (90) days from the date of original purchase. All implied warranties are limited in duration to this ninety (90) day period. THEREAFTER, ANY IMPLIED WARRANTIES, INCLUDING IMPLIED WARRANTIES OF MERCHANTABILITY AND FITNESS FOR A PARTICULAR PURPOSE ARE EXCLUDED. THIS WARRANTY IS IN LIEU OF ALL OTHER WARRANTIES, WHETHER ORAL OR WRITTEN, EXPRESSED OR IMPLIED.

If you believe the media is defective, please return it during the ninety day period to the address shown below. A defective diskette will be replaced without charge provided that it has not been subjected to misuse or damage.

This warranty does not extend to the software or information recorded on the media. The software and information are provided "AS IS." Any statements made about the utility of the software or information are not to be considered as express or implied warranties. Autodesk Press will not be liable for incidental or consequential damages of any kind incurred by you, the consumer, or any other user.

Some states do not allow the exclusion or limitation of incidental or consequential damages, or limitations on the duration of implied warranties, so the above limitation or exclusion may not apply to you. This warranty gives you specific legal rights, and you may also have other rights which vary from state to state. Address all correspondence to:

Autodesk Press
3 Columbia Circle
P. O. Box 15015
Albany, NY 12212-5015